NORTHEASTERN ILLINOIS UNIVERSITY
3 1224 00423 5684

D1223356

AN ESSAY IN UNIVERSAL SEMANTICS

TOPOI LIBRARY

VOLUME 1

Topoi Library *is sponsored by the Department of Philosophy and the School of Humanities at the University of California, Irvine*

ACHILLE C. VARZI

Department of Philosophy,
Columbia University, New York

AN ESSAY IN UNIVERSAL SEMANTICS

KLUWER ACADEMIC PUBLISHERS

DORDRECHT / BOSTON / LONDON

A C.I.P. Catalogue record for this book is available from the Library of Congress.

ISBN 0-7923-5629-2

Published by Kluwer Academic Publishers,
P.O. Box 17, 3300 AA Dordrecht, The Netherlands.

Sold and distributed in North, Central and South America
by Kluwer Academic Publishers,
101 Philip Drive, Norwell, MA 02061, U.S.A.

In all other countries, sold and distributed
by Kluwer Academic Publishers,
P.O. Box 322, 3300 AH Dordrecht, The Netherlands.

Printed on acid-free paper

Printed in the Netherlands.

for my parents

EDITORIAL PREFACE

Like the journal *TOPOI,* the *TOPOI Library* is based on the assumption that philosophy is a lively, provocative, delightful activity, which constantly challenges our inherited habits, painstakingly elaborates on how things could be different, in other stories, in counterfactual situations, in alternative possible worlds. Whatever its ideology, whether with the intent of uncovering a truer structure of reality or of shooting our anxiety, of exposing myths or of following them through, the outcome of philosophical activity is always the destabilizing, unsettling generation of doubts, of objections, of criticisms.

It follows that this activity is intrinsically a *dialogue,* that philosophy is first and foremost philosophical discussion, that it requires bringing out conflicting points of view, paying careful, sympathetic attention to their structure, and using this dialectic to articulate one's approach, to make it richer, more thoughtful, more open to variation and play. And it follows that the spirit which one brings to this activity must be one of tolerance, of always suspecting one's own blindness and consequently looking with unbiased eye in every corner, without fearing to pass a (fallible) judgment on what is there but also without failing to show interest and respect.

It is no rhetoric then to say that the *TOPOI Library* has no affiliation to any philosophical school or jargon, that its only policy is to publish exciting, original, carefully reasoned works, and that its main ambition is to generate serious and responsible exchanges among different traditions, to have disparate intellectual tools encounter and cross-fertilize each other, to contribute not so much to the notarization of yesterday's syntheses but rather to the blossoming of tomorrow's.

Contents

Acknowledgments

This essay originated as my doctoral dissertation, which was submitted in 1994 to the Department of Philosophy in the University of Toronto. Much that appears here appeared there, but the rewriting has been extensive and some parts have been substantially revised. I have also included new material drawn from recent work in the area, and I have made an effort to do away with outdated conjectures and speculations.

Some very special people helped me during this project, and I would like to express my deep gratitude to them: Hans Herzberger, my teacher, whose words and letters have been more supportive than I can say; Mark Vorobej, whose encouragement at crucial junctures proved invaluable; Piero Bonatti, Roberto Casati, Alex Simpson, and Jamie Tappenden, who helped me to clean up my ideas over the years; and Friederike Oursin, my wife, who shared every single stage of this work. I would also like to thank Peter Apostoli, John Bell, Hector Levesque, and Alasdair Urquhart for penetrating comments on my dissertation, some of which have helped to bring this monograph to its present form. During the final rewriting, I have benefited also from discussions with Horacio Àrlo Costa, John Collins, Haim Gaifman, Dominic Hyde, Isaac Levi, Graham Priest, and the participants in various talks that I have had the opportunity to give in various places; I am thankful to all for their remarks and provocations. Finally, I would like to thank Ermanno Bencivenga, both for his support and encouragement during the very final stage, and for the influence that some of his work has had on my own way of thinking about these matters from the beginning.

INTRODUCTION

Semantics, understood as a formal theory of the relationships between languages and their models, must involve some means of describing these domains: some notion of a language, and some notion of a model. To the extent that the account provides an accurate construal of both notions, to that extent it properly qualifies as "semantics". And to the extent that it does not depend on specific constraints upon either notion, to that extent the account is "universal", i.e., philosophically neutral and widely applicable.

It is this general approach that sets the background, and somehow the foreground, of the present essay. On the one hand, it seems to me that a broad outlook can still be beneficial to our understanding of a number of semantic issues, and this can only be achieved by working with a suitably wide notion of a language. A common limitation of much work in model-theoretic semantics—including "non-standard" theories—seems to me to lie in its relative idiosyncrasy. With few notable exceptions, proposals have been local rather than taking place within global linguistic frameworks. And although the resulting variety of theories has a corresponding variety of merits, how to assess them and to look for further progress is often subject to the possibility of providing more extensive accounts of the relevant linguistic structures. On the other hand, I am also convinced of the importance of working with a widest possible notion of a model, free of various conditions that are typically assumed in the context of formal semantics. A severe limitation of most theories— especially "standard" ones—is their commitment to some view or other as to what is to count as an "admissible" model for a given language. And although this way of proceeding has proven to be convenient for special purposes, all the same a certain embarrassment arises when one has to resort to artificial measures to regiment alternative intuitions or intractable phenomena.

My anxiety about a general notion of language is chiefly methodological. However, the need for a general notion of model is more substantive and arguably central to any semantic project that aims at some universality. For instance, a major requirement of classical semantics is that every model be consistent (or coherent) as well as complete (or determinate) relative to the given language. Broadly speaking, this means that in classical semantics all models are assumed to provide a homomorphic interpretation of the language. They are made of well-defined, clear-cut entities, linked to one another and to the language's expressions in a univocal way: all singular terms (proper names and the like) are assumed to have a unique denotation; all general terms (predicates and the like) must represent properties or relations that are uniquely defined for each object in the domain; all sentences must receive a unique truth-value; and so on. Now, such an approach is perfectly defensible and it may well be advantageous for a number of special purposes. (It is certainly advantageous if we are interested in the semantics of mathematical theories, for instance.) However, when it comes to the general case the burden of the defense is philosophical—not semantical. There is no *a priori* semantic reason to rule out the possibility that (our representation of) what a language is about involves "gaps" and "gluts" of sorts. Indeed, since there is no general criterion for sieving out troublous forms of incompleteness from innocuous ones, there is no general and effective way of ruling out gaps without discarding many unproblematic cases as well. And since there is no general antidote against inconsistency, there is no guarantee that gluts can be ruled out without also rendering a great deal of perfectly innocent semantic thinking impossible.

This theoretical anxiety goes hand in hand with more practical considerations. A model represents the world, or one way in which the world could be. But a representation may be incomplete, or even inconsistent, and yet perfectly adequate for most purposes. Data bases and works of fiction are models of this sort. A data base need not be able to answer every possible query that can be formulated in the query language. And a novel need not be based on a thorough description of its imaginary world to make sense. We are only told a few things about the world of Sherlock Holmes, yet we can understand the Holmes stories and reason about them. As it turns out, we are also told inconsistent things: Watson's war wound is supposed to be in his shoulder, but also in his leg. Yet one discrepancy is no logical chaos. Lots of facts are perfectly clear about Holmes and Watson, in spite of the gaps and the gluts involved in the model of their world. These sorts of situation are quite normal and may arise in any model. It seems to me that a good semantic theory should be able to handle them.

There is, moreover, a general and widespread concern that a semantic theory based on the requirement of a perfect homomorphism between languages and models will be inapplicable to ordinary language. Vagueness, ambiguity, open texture, presupposition failure, sortal incorrectness—these are all common features of the language we ordinarily speak (but also the language of the exact sciences and the language of philosophy) that seem at odds with the requirements of standard semantics. For each of these "problems"—and for many others indeed—there is by now one or more corresponding "solutions" within some more or less standard framework. However, the resulting accounts are admittedly severe—sometimes *ad hoc*, sometimes ingenious, but rarely natural, and there is a growing consensus that a more liberal attitude towards those recalcitrant phenomena could yield a better overall theory. At any rate, it seems to me that even a successful regimentation of these phenomena would not diminish the value of semantic theories capable of taking them at face value—theories in which those "problems" are part of the data to be taken into account, rather than accidents to be explained away.

With all this, the last three decades have featured a great deal of work under the rubric of "going general" in model-theoretic semantics, sometimes with excellent results. Especially in the areas of natural language semantics and philosophical logic (but also in computer science and certain branches of mathematical logic), there has been a remarkable outgrowth of new semantic theories and frameworks. However, most of this work has arisen out of specific needs and for specific purposes—for instance, to account for one or another of the linguistic phenomena mentioned above, or to establish a completeness result for some logical calculus. Not much work has been done in the direction of a "general" generalization. Also, gaps have played a leading role in this trend: gluts have been ostracized for a much longer time, and the intrinsic duality between the two notions has not been given much consideration. This essay is an attempt to overcome these limitations. My approach will be top-down rather than bottom-up. (This is why I want to start with a reasonably abstract notion of language.) And the upshot will be a framework within which a large variety of semantic theories naturally fall and from which the semantics of a wide class of logics (standard and nonstandard) may be obtained as special cases. There is no claim to complete universality, of course. But my hope is that the approach, if not the framework as a whole, may shed new light on some important issues in formal semantics.

The details of the framework form the core of the first chapter, *Foundations*. I first define languages and models, and then suggest a general technique for constructing semantic evaluations (of a language $\mathcal{L}$ on a model $\mathcal{M}$). Very roughly, the main idea is grounded on some structural properties of the class of all models and may be viewed as implementing a form of "supervaluationism": the value of an expression relative to a given model is a function of the values of that expression relative to various ways of "sharpening" the model (by filling in its gaps and weeding out its gluts, if any). The working of this account is illustrated in connection with some examples of increasing complexity. In the second chapter, *Developments*, I turn to the main properties of the resulting apparatus. Here again the emphasis is on the general picture; but I am also concerned with a study of the specific conditions under which some major results of standard semantics can be extended to our more general account. While this concern might at times appear scholastic, I hope it will help to provide a better understanding of the strengths and weaknesses of the approach pursued here, as well as offer new insights into why those results hold in the standard case. In particular, it is shown that the notion of logical validity as applied to single expressions is rather standard, and in fact reduces to the corresponding classical notion when this is available. Thus, gaps and gluts can be explained away as local phenomena that do not metastasize throughout the framework. However, when it comes to the general notion of logical entailment the picture is more intricate, and the notion of argument validity does not appear to be recursively specifiable except in very special cases. To shed light on this picture, a cost-benefit analysis of the framework—along with some general philosophical reflections—is offered in the *Concluding Remarks*.

Just about every technical notion in the following is construed, or assumed to be construed, in set-theoretic terms. The precise content of the underlying system of set theory will only rarely be of any importance, and I shall make little effort to exhibit the details of my discourse within it. At places, however, some explicit commitment on such matters is wanted, so I have included an *Appendix* which sorts these things out and also explains all necessary notational and terminological conventions. This should make the exposition self-contained. Certain passages, such as the proofs of the basic facts and examples in Section·1.3 [in square brackets], may be omitted by those who are prepared to take them on trust. In the main, however, the conceptual apparatus is intended as a whole and frequent use will be made at later stages of material introduced earlier.

1. FOUNDATIONS

I shall set up the basic framework in three steps. First, in Section 1.1, I shall introduce the notion of a *language*, regarded as a purely combinatorial, uninterpreted system. Next, in Section 1.2, I shall turn to the relative notion of a *model*, or interpretation system: this notion will be characterized in a parallel fashion and will allow for the kind of generality which I am aiming at. Finally, in Section 1.3, I shall turn to the task of bridging languages and models via the concept of a *valuation*: this is the most problematic and controversial part but also—I hope—the most interesting one.

1.1. LANGUAGES

I shall not be concerned with any specific language, but only with the bone structure of the concept of a language insofar as this matters for the purpose of clarifying basic semantic notions. In this sense my characterization will be minimalistic: a language is essentially a system of (well-formed) expressions generated from a (well-ordered) class of symbols by means of some (well-grounded) structural operation.

1.1.1. PRELIMINARIES

To spell out the details of the definition in a convenient form (avoiding all discussion concerning the exact nature of a language's symbols and structural operation), we can make use of the auxiliary concept of a *type*, or *category index*. That is, we can suppose that the symbols of any language form a family of objects (of some sort) indexed by a fixed set $\mathcal{T}$ of types, and we may refer to such an indexing to determine how those objects may combine with one another to produce what is to count as an expression of the language. Besides ontological neutrality and conceptual simplicity, this procedure will also have some derivative advantages, which will become clearer as the discussion proceeds.[1]

The set of all types, I shall assume, is built up recursively from an initial stock of *primitive* (or *individual*) *types*. Intuitively, such types are to be associated with those categories of expressions whose grammatical status need not—or cannot—be analyzed in terms of other categories. One of these could be, for example, the category Declarative Sentence; another could be the category Proper Name. In fact, for most purposes these two basic categories would probably suffice, although someone would perhaps insist on adding a third one, Common Noun, or on dispensing with Proper Name in favor of an alternative basic category Noun Phrase, or Verb Phrase.[2] In general, however, I shall make little effort here to select such primitives with care. What we need to do is simply to assume a certain number of primitive, undefined types. And to allow for the greatest generality, it seems convenient to start off with a potentially infinite number, bypassing the problem of specifying the intuitive status of each of them. Thus, for definiteness I stipulate to identify the primitive types with the natural numbers

$$0, 1, 2, \ldots$$

taking 0 and 1 to represent the traditionally fundamental categories of declarative sentences and proper names, respectively. (Nothing of what follows depends significantly on this specific choice, or on the fact that

---

[1] The syntactic framework developed here is in line with the general approach of categorial grammar, with some formal simplifications leaning on related insights in combinatory type theory. For the former, the classical references are Leśniewski [1929], Carnap [1934], and Ajdukiewicz [1935]. For the latter, the seminal contributions date back to Schönfinkel [1924] and Curry [1929, 1930], though the type hierarchy (the analogue of our $\mathcal{T}$) was not introduced until Curry [1953] (see Sanchis [1964]). The two theories, in turn, have been developed under the impact of Husserl [1900-01] and Russell [1908], respectively. For a general outline, see Casadio [1988] and Wood [1994]; for a survey of connections between the two disciplines, as they arise out of recent developments, see van Benthem [1990] and Steedman [1988, 1993]; for an overview of the use of types in semantic analysis, see Carpenter [1996] and Turner [1997].

[2] See, for instance, Lewis [1970] and Cresswell [1973].

the set ω of all natural numbers is denumerable: all that matters is that the class of primitive types be isomorphic to some fixed ordinal α. Even the proviso that $\alpha \geq 2$ is dictated merely by reasons of convenience and common practice.)

On this basis, an infinite set of *derived* (or *functional*) *types* is obtained by introducing some operation on the set of primitive types, say the operation $\langle\ \rangle$ of pair formation. More generally, I shall assume that whenever t_1 and t_2 are any types, primitive or derived, a new derived type may be formed, which we may identify with the ordered pair

$$\langle t_1,t_2\rangle.$$

The intuitive idea is to think of such a type as corresponding to those categories of expressions (functors, in the traditional terminology) that produce expressions of type t_2 when combined with expressions of type t_1. Thus, for instance, if 0 and 1 are interpreted as above, then $\langle 0,0\rangle$, $\langle 0,1\rangle$, $\langle 1,0\rangle$, and $\langle 1,1\rangle$ will be the types of such categories of functors traditionally referred to as (monadic) Connectives, Subnectives, Predicates, and Operators, respectively; $\langle\langle 0,0\rangle,\langle 0,0\rangle\rangle$, $\langle\langle 0,1\rangle,\langle 0,1\rangle\rangle$, etc., will be the types of the corresponding (monadic) Modifiers; and so on. Here are some examples from English:

it is not the case that	connective	$\langle 0,0\rangle$
the claim that	subnective	$\langle 0,1\rangle$
the husband of	operator	$\langle 1,1\rangle$
cries	predicate	$\langle 1,0\rangle$
is smarter than	relational predicate	$\langle 1,\langle 1,0\rangle\rangle$
loudly	predicate modifier	$\langle\langle 1,0\rangle,\langle 1,0\rangle\rangle$

Note indeed that if we introduce derived types by means of a binary operation, $\langle\ \rangle$, we can initially speak of "monadic" functors only. However, this implies no loss of generality, for every functor can eventually be regarded as a monadic functor of a certain type. The relational predicate 'is smarter than' illustrates this point: 'is smarter than' is naturally regarded as a dyadic functor that makes a sentence out of two proper names (as in 'Ann is smarter than Bruce'); but it can equally be regarded as a monadic functor which, when applied to any given proper name (say 'Bruce'), produces an expression ('is smarter than Bruce') that behaves again as a monadic functor: a functor that makes a sentence out of a proper name ('Ann'). We can therefore treat such a relational predicate as having type $\langle 1,\langle 1,0\rangle\rangle$. Likewise, dyadic connectives, subnectives, operators, etc. may be thought of as having the types $\langle 0,\langle 0,0\rangle\rangle$, $\langle 0,\langle 0,1\rangle\rangle$, $\langle 1,\langle 1,1\rangle\rangle$, etc. More generally, whenever $n \geq 1$ and $t_1, \ldots, t_n, t_{n+1}$ are types, primitive or derived, the type

$$\langle t_1, \langle t_2, \langle \ldots, \langle t_n, t_{n+1}\rangle \ldots \rangle\rangle\rangle$$

may be used to represent the category of those "n-adic" functors that combine with n-tuples of expressions of type $t_1, t_2, \ldots, t_n$ (in this order) to produce expressions of type t_{n+1}. In this sense, as long as the second coordinate of a derived type is allowed to be a derived type itself, it will appear that our relying on $\langle\ \rangle$ imposes no significant restriction on the class of possible functors.[3] The only important thing is that each derived type be distinct from all primitive types, and that no derived type be generated in more than one way. (In other words, the set of all types must be freely generated from the set of primitive types.)

On the intended interpretation, the classification of all types into two main sorts, individual and functional, reflects of course the traditional tenet that the linguistic combination of constituents into constitutes should be regarded not as a concatenation *inter pares*, but rather as the result of the application of one constituent upon the other(s). Thus, in general, two expressions x and y may combine to produce a new expression z (of a certain type) if and only if x is a functor and y an expression whose type coincides with the first coordinate of the type of x. More specifically, the assumption is that the expressions of any lan-

---

[3] The idea goes back to Schönfinkel [1924]. Of course, we could equally well follow the opposite strategy, allowing for n+2-adic derived types for each $n \in \omega$ while requiring their second coordinate to be a basic type. On the intended interpretation, the equivalence of the two procedures corresponds to the set-theoretic isomorphism $A^{B_1 \times B_2 \times \ldots \times B_{n+1}} \approx (\ldots((A^{B_1})^{B_2})\ldots)^{B_{n+1}}$ (but see infra, note 18). Various authors, following in the footsteps of Bar-Hillel [1950], Lambek [1958, 1961], and Geach [1970], have also considered allowing for greater flexibility in the rules of type assignment or functional application (see Bach [1984] and Wood [1994] for overviews; see also Ranta [1994] for extensions obtained by basing the grammar on a constructive type theory, in the form developed by Martin-Löf [1984]): although such extensions are in the spirit of further generality, I believe the simpler framework adopted here is sufficiently powerful to induce no dramatic conceptual limitations.

guage $\mathcal{L}$ can be recursively specified on the basis of some assignment of types to the symbols of $\mathcal{L}$: for each type t, the expressions of $\mathcal{L}$ will include a certain (possibly empty) set of expressions of type t, and this will comprise all the symbols initially assigned to t plus all those expressions that result from applying a given structural operation (usually some form of concatenation or juxtaposition) to pairs of expressions of type $\langle t',t\rangle$ and t', respectively.

Note that, for the reasons explained above, one only needs a *binary* operation to generate expressions in this way.[4] Not so obvious perhaps is the fact that one would not achieve greater generality if one allowed expressions to be built up by means of many different modes of formation. For instance, languages involving formal variables or variable-binders (such as integrals, quantifiers, and the like) seem to run afoul of this simple apparatus, unless a structural operation of functional abstraction is assumed in addition to the operation of functional application described here.[5] More generally, it might be suggested that functional abstraction is needed in order to bring out certain important relations between different levels of linguistic analysis, for instance, between deep logical structure and surface realizations.[6] I don't know whether this is indeed so. But the examples given below will show that one can go a long way (including dealing with various forms of variable binding) without such extensions. Thus, also in this respect it will appear that the present approach—though obviously very simplistic and idealized—imposes no major restriction on the class of admissible languages.

---

[4] By contrast, if we followed the alternative procedure mentioned in note 3, we would need an n+2-ary structural operation whenever $\mathcal{L}$ involves "n+2-adic" functors, at the price of considerable complexities in the overall apparatus.

[5] This view can be traced back to Ajdukiewicz [1935], Part 2, and is partly supported by a result of Bar-Hillel, Gaifman & Shamir [1960] (which says that categorial grammars based exclusively on functional application are equivalent to context-free phrase-structure grammars). See for instance Marsh & Partee [1987]. For an overview of the issues, see Jacobson [1996], § 5. Some arguments in support of the simpler approach advocated here may be found in Varzi [1993].

[6] Cresswell [1977] conjectures that all "semantically significant" transformational derivations might be seen as sequences of lambda-conversions, in the sense of Church [1940, 1941]. Alternatively, some authors have proposed supplementing functional application with a set of transformation-like rules: this was the gist of Lewis [1970] or Partee [1975, 1976].

Let then $\mathcal{T}$ be the set of all types, primitive or derived, built up in the indicated way:

$$\mathcal{T} = \text{the closure of } \omega \text{ under } \langle\,\rangle.$$

Keeping in mind the intuitive roles described above, the general definition of a language goes as follows. (I shall first give the definition, and then illustrate its working with some examples.)

1.1.2. DEFINITION

A *language* is an ordered triple $(\mathbf{s},\mathbf{g},\mathbf{E})$ satisfying the following general conditions:

(a) $\mathbf{s}$ is a one-one function such that $\mathcal{D}\mathbf{s}: \alpha \to \mathcal{T}$ for some ordinal α;
(b) $\mathbf{g}$ is a one-one function such that $\mathcal{R}\mathbf{g} \cap \mathcal{R}\mathbf{s} = \varnothing$ (and $\mathcal{D}\mathbf{g}$ as below);
(c) $\mathbf{E}$ is the smallest system of sets closed under the properties: (i) if $\langle\beta,t\rangle \in \mathcal{D}\mathbf{s}$, then $t \in \mathcal{D}\mathbf{E}$ and $\mathbf{s}(\beta,t) \in \mathbf{E}_t$; and (ii) if $t,t' \in \mathcal{D}\mathbf{E}$, $x \in \mathbf{E}_t$, $y \in \mathbf{E}_{t'}$, and $t = \langle t',t''\rangle$, then $t'' \in \mathcal{D}\mathbf{E}$ and $\mathbf{g}(x,y) \in \mathbf{E}_{t''}$.

(For definiteness, I shall assume that $\mathbf{g}$ is the smallest function satisfying the stated conditions, i.e., $\mathbf{g}(x,y)$ is defined only if $\mathbf{g}(x,y) \in \bigcup\mathcal{R}\mathbf{E}$ by clause (c).)

It is understood that if $\mathcal{L} = (\mathbf{s},\mathbf{g},\mathbf{E})$ is any language, then the elements of $\mathcal{R}\mathbf{s}$ are the *symbols* of $\mathcal{L}$ and $\bigcup\mathcal{R}\mathbf{E}$ is the class of all *expressions* generated from those symbols by repeated application of the *structural operation* $\mathbf{g}$. The requirement that $\mathbf{s}$ and $\mathbf{g}$ be one-one functions is to avoid ambiguities; combined with the requirement that $\mathbf{g}$ be well-grounded on $\mathcal{R}\mathbf{s}$ (i.e., $\mathcal{R}\mathbf{g} \cap \mathcal{R}\mathbf{s} = \varnothing$), this will secure that every expression of $\mathcal{L}$ can be uniquely represented either as a symbol or as a compound of the form $\mathbf{g}(x,y)$. In particular, by an *$\mathcal{L}$-symbol of type t* (where $t \in \mathcal{R}\mathcal{D}\mathbf{s}$) I shall understand any element $x \in \mathcal{R}\mathbf{s}$ such that $x = \mathbf{s}(\beta,t)$ for some $\beta \in \mathcal{D}\mathcal{D}\mathbf{s}$ (such a β will simply be called the *alphabetic* or *ordinal index* of x); and an *$\mathcal{L}$-expression of type t* (where $t \in \mathcal{D}\mathbf{E}$) will be any element of $\mathbf{E}_t$. Since $\mathcal{D}\mathbf{s}$ is a function, this assignment of types to symbols and therefore to expressions is sure to be unique: the sets in the system $\mathbf{E}$ are all pairwise disjoint. (This follows immediately by induction on the number of applications of $\mathbf{g}$.)

1.1.3. REMARKS

In the following, the class of all languages defined in 1.1.2 will be denoted by '$\mathbb{L}$'. I shall use the letter '$\mathcal{L}$', possibly with subscripts or superscripts, exclusively as a variable on $\mathbb{L}$, and I shall conform to the notational convention that $\mathcal{L}=(\mathbf{s},\mathbf{g},\mathbf{E})$, $\mathcal{L}'=(\mathbf{s}',\mathbf{g}',\mathbf{E}')$, $\mathcal{L}_j=(\mathbf{s}_j,\mathbf{g}_j,\mathbf{E}_j)$, and so on. There is some ambiguity in this convention, due to the fact that '$\mathbf{E}_j$' can denote either the third component of a language $\mathcal{L}_j$, or a basic category of a language $\mathcal{L}$. The context, however, should always prevent confusion.

There is no doubt that $\mathbb{L}$ contains some peculiar elements, which will hardly play any role in our discussion. For example, clause (a) in the definition allows for the possibility that a language $\mathcal{L}\in\mathbb{L}$ involve an arbitrary number of symbols for each type $t\in\mathcal{T}$, but it is clear that unless some $\mathcal{L}$-symbols are assigned functional types, the structural operation of $\mathcal{L}$ would be empty and no compound expression could be generated. Also, there is nothing in the definition to secure that all symbols of a given language can actually be used to produce expressions of type 0, i.e., sentences. In fact, most elements of $\mathbb{L}$ involve no sentences at all. Such oddities, however, are irrelevant for our purposes, and I shall therefore refrain from introducing unnecessary complexities into the account. In any case, note that the definition is neutral with respect to both the ontological status of a language's symbols (the range of $\mathbf{s}$) and the outputs of the structural operation (the range of $\mathbf{g}$). One can think of the symbols as marbles, sets of tokens, expressions of some other language (e.g., English), and one can think of expressions as sequences of marbles, sets of sequences of tokens, well-formed expressions of some other language (English). It is not even assumed that a compound expression actually *consists* of its component expressions, i.e., of the elements used to construct it: $\mathbf{g}$ may incorporate all sorts of idiosyncratic transformations, so that, for instance, if x is the English phrase 'it is not the case that' and y is 'Albert cries', the expression $\mathbf{g}(x,y)$ may be the English sentence 'Albert does not cry'. None of this is crucial. But it helps disentangling what is semantically relevant from what is not.

In terms of the above notions, we can then introduce some additional terminology and notation that will prove useful as the discussion proceeds. Thus, if $\mathcal{L}$ is any language in $\mathbb{L}$, I shall generally write

'$SYM_{\mathcal{L}}$' instead of '$\mathcal{R}$s' and '$EXP_{\mathcal{L}}$' instead of '$\bigcup\mathcal{R}$E', and I may sometimes speak of an element $x \in EXP_{\mathcal{L}}$ as of an *individual* or *functional* expression of $\mathcal{L}$ (in short $x \in IND_{\mathcal{L}}$ or $x \in FUN_{\mathcal{L}}$) depending on whether the type of x is an individual or a functional type, respectively. Accordingly, $IND_{\mathcal{L}} \cap FUN_{\mathcal{L}} = \varnothing$ and $IND_{\mathcal{L}} \cup FUN_{\mathcal{L}} = EXP_{\mathcal{L}}$. (Clause (c) in the definition does not leave room for **g** to have self-applicative arguments, i.e., there is no x for which $\mathbf{g}(x,x)$ is defined. One could relax this condition, but I shall not consider that possibility here.[7])

Given an expression $x \in EXP_{\mathcal{L}}$, we can then define its *length* (relative to $\mathcal{L}$) to be the number of times that **g** needs to be applied in order to generate it, in symbols $\lambda_{\mathcal{L}}(x)$:

if $x \in SYM_{\mathcal{L}}$, then $\lambda_{\mathcal{L}}(x)=0$
if $x=\mathbf{g}(y,z)$, then $\lambda_{\mathcal{L}}(x)=1+\lambda_{\mathcal{L}}(y)+\lambda_{\mathcal{L}}(z)$.

Similarly, the *width* of x (relative to $\mathcal{L}$) will be the set $\omega_{\mathcal{L}}(x)$ of its constituents (relative to **g**):

if $x \in SYM_{\mathcal{L}}$, then $\omega_{\mathcal{L}}(x)=\{x\}$
if $x=\mathbf{g}(y,z)$, then $\omega_{\mathcal{L}}(x)=\{x\} \cup \omega_{\mathcal{L}}(y) \cup \omega_{\mathcal{L}}(z)$.

In addition, we may write '$\tau_{\mathcal{L}}(x)$' to denote the type of x (relative to $\mathcal{L}$) and then classify x in relation to its *rank* and *degree*: these values can be defined as the finite ordinals

$\rho_{\mathcal{L}}(x)=r(\tau_{\mathcal{L}}(x))$
$\delta_{\mathcal{L}}(x)=d(\tau_{\mathcal{L}}(x))$

respectively, where r and d are the two smallest functions on $\mathcal{T}$ with the following properties:

if $t \in \omega$, then $r(t)=d(t)=0$
if $t=\langle t_1,t_2\rangle$, then $r(t)=1+r(t_1)$ and $d(t)=1+d(t_2)$.

Due to the above-mentioned conditions on **s** and **g**, it is easily verified that all these notions are well defined, i.e., $\lambda_{\mathcal{L}}$, $\omega_{\mathcal{L}}$, $\tau_{\mathcal{L}}$, $\rho_{\mathcal{L}}$, and $\delta_{\mathcal{L}}$ are all functions. However, let me stress that although we have $\lambda_{\mathcal{L}}(x)=\varnothing$ and

---

[7] This was implicit in my way of introducing the type hierarchy. For a natural language oriented investigation of type-free constructions, see Kamareddine & Klein [1993].

$\omega_{\mathcal{L}}(x)=\{x\}$ for all $x\in SYM_{\mathcal{L}}$, it is by no means implied that a language's symbols must lack any internal structure, but only that this structure—if any—is irrelevant to the syntax of the language.[8] (Examples of structured symbols will actually be considered below, in connection with the treatment of variable binding.) Also, the fact that we can only speak of *the* type of a symbol, and consequently of an expression (as reflected by the function $\tau_{\mathcal{L}}$), is only a seeming limitation: although it is not difficult to imagine words belonging to more than one grammatical category, we can always deal with such cases by treating them as distinct symbols with a common "surface realization". (For example, an English word like 'light' may seem to function as either a noun or an adjectival modifier, but one can also deal with this peculiarity the way common dictionaries do: by distinguishing between a word 'light$_1$', to be classified as a noun (arguably of type $\langle 1,0\rangle$), and a word 'light$_2$', to be classified as a modifier (of type $\langle\langle 1,0\rangle,\langle 1,0\rangle\rangle$).)

Finally, the definition of $\rho_{\mathcal{L}}$ and $\delta_{\mathcal{L}}$ should now make clear the sense in which relying on a binary operation $\langle\,\rangle$ does not yield any significant restriction on the class of admissible functors: a language's vocabulary may in fact involve symbols and expressions of infinitely many ranks and degrees (only infinite ranks and degrees are ruled out, due to the intended interpretation of functional types). Notationally, this can be made explicit, for instance, by agreeing that whenever $x\in \mathbf{E}_{\langle t_1,\langle t_2,\langle\ldots,\langle t_n,t_0\rangle\ldots\rangle\rangle\rangle}$ and $y_1\in \mathbf{E}_{t_1},\ldots,y_n\in \mathbf{E}_{t_n}$, the expression

$$\mathbf{g}(\ldots(\mathbf{g}(\mathbf{g}(x,y_1),y_2),\ldots),y_n)$$

may be written as if x were a true n-adic functor,

$$\mathbf{g}(x,y_1,\ldots,y_n),$$

or simply as

$$x(y_1,\ldots,y_n).$$

It is understood that the foregoing applies to every language $\mathcal{L}\in\mathbb{L}$. Further conventions might then be introduced when a particular language or class of languages is under consideration. For example, if the

---

[8] It might be more appropriate to speak of "morphemes" (Lyons [1969], § 5.3).

expressions of a language $\mathcal{L}$ are construed as finite sequences of symbols and **g** as the operation of sequence formation, or concatenation (so that $\mathbf{g}(x,y)=x^\frown y$ for x and y as appropriate), then the length of a compound expression $z\in EXP_{\mathcal{L}}$ can simply be defined as $\mathcal{D}z-1$, and one could speak of the n-th term of the sequence as of the *n-th symbol* of z.

Indeed, the generality of definition 1.1.2 lies precisely in the fact that it allows one to single out any desired language or class of languages simply by supplementing clauses (a)–(c) with the appropriate additional conditions. Thus, again, one may want to impose as a desirable requirement that the functional expressions of a language $\mathcal{L}$ always yield individual expressions, in the sense that whenever $\mathbf{E}_{\langle t',t\rangle}$ is defined (for any $t,t'\in\mathcal{T}$), $\mathbf{E}_{t'}$ is also defined, and hence so is $\mathbf{E}_t$. By induction this implies that whenever $\mathbf{E}_{\langle t',t\rangle}$ is defined, there is a non-empty category $\mathbf{E}_{k(t)}$ such that

if $t\in\omega$, then $k(t)=t$,
if $t=\langle t_1,t_2\rangle$, then $k(t)=k(t_2)$.

And further, among such languages one may want to pick only those languages whose categories of expressions always "cancel" to a given type t, i.e., can be used to generate expressions of type t: in short, only languages $\mathcal{L}\in\mathbb{L}$ such that

$$EXP_{\mathcal{L}}=\bigcup\{\omega_{\mathcal{L}}(x): x\in\mathbf{E}_t\}.$$

As already remarked, most languages of the sort usually discussed in logical theory can formally be regarded as languages whose categories of expressions always cancel to 0, and in the following developments such languages will be studied in greater detail. As a rule, however, the basic apparatus will be spelled out for the general case.

1.1.4. EXAMPLES

Here are three examples of how familiar languages may be retrieved as special cases of the notion introduced in 1.1.2. Let $\mathcal{L}$ be any language in $\mathbb{L}$; and let us agree to write '$\langle t^{\langle n\rangle},t'\rangle$' as a simplified notation for '$\langle t_0,\langle t_1,\langle\ldots,\langle t_{n-1},t'\rangle\ldots\rangle\rangle\rangle$' whenever $n\in\omega$, $t,t'\in\mathcal{T}$, and $t_i=t$ for each $i\in n$ (in case n=0, $\langle t^{\langle n\rangle},t'\rangle$ reduces simply to t'). Then:

(A) $\mathcal{L}$ is a *sentential language* if and only if $\mathcal{RD}$**s**$\subseteq\{\langle 0^{\langle n\rangle},0\rangle: n\geq 0\}$, i.e., if and only if $SYM_{\mathcal{L}}$ comprises a certain stock of *sentence symbols* (symbols of type 0) along with some *connectives* (of type $\langle 0^{\langle n\rangle},0\rangle$, for various $n>0$).

To introduce some familiar notation, suppose $\mathcal{L}$ is a sentential language whose symbols include a dyadic connective $\downarrow$. Then a sentence of the form

$$\mathbf{g}(\mathbf{g}(\downarrow,A),B)$$

(where $A,B\in \mathbf{E}_0$) may naturally be written as

$$(A\downarrow B).$$

If **g** is the operation of concatenation, as considered above, this is just another name of what is, in effect, a bracket-free sentence '$\downarrow^\frown A^\frown B$'. On the other hand, one could as well construe **g** so that the actual sentence (the value of **g** for the arguments $\mathbf{g}(\downarrow,A)$ and B) is the string '$(A\downarrow B)$': in that case the metalinguistic notation would mirror the linguistic object. (This is possible even if parentheses are not among the language's symbols, since expressions need not be made up *of* symbols only.)

More familiar notation may be introduced by contextual definitions, as the case may be. For instance, suppose $\downarrow$ is the only connective of $\mathcal{L}$, understood intuitively as the *joint denial* connective 'neither . . . nor'. Then the following corresponds to the customary notation for *negation, disjunction, conjunction, material conditional*, and *material biconditional*, respectively:

(a) '$(\neg A)$' for '$(A\downarrow A)$'
(b) '$(A\vee B)$' for '$(\neg(A\downarrow B))$'
(c) '$(A\wedge B)$' for '$((\neg A)\vee(\neg B))$'
(d) '$(A\rightarrow B)$' for '$((\neg A)\vee B)$'
(e) '$(A\leftrightarrow B)$' for '$((A\rightarrow B)\wedge(B\rightarrow A))$'

We shall see below how this interpretation can be characterized semantically.[9] (Of course, one can also have sentential languages containing any

---

[9] See examples 1.2.4(A) and 1.3.4(A). On the intended interpretation, definitions (a)-(e) are those of Sheffer [1913] (also in Peirce [1880]).

number of such connectives among their actual symbols. The basic format would be the same in each case.)

(B) $\mathcal{L}$ is an *elementary language* if and only if $\mathcal{RD}s \subseteq \{\langle t^{\langle n \rangle}, t' \rangle : n \geq 0$ and $t, t' \leq 1\}$, i.e., if and only if $IND_{\mathcal{L}}$ comprises a set $SENT_{\mathcal{L}}$ of *sentence symbols* (of type 0) and a set $NAME_{\mathcal{L}}$ of *name symbols* (of type 1) while $FUN_{\mathcal{L}}$ comprises, for each $n > 0$, a set $CON_{\mathcal{L},n}$ of *n-ary connectives* (of type $\langle 0^{\langle n \rangle}, 0 \rangle$), a set $SUB_{\mathcal{L},n}$ of *n-ary subnectives* (of type $\langle 0^{\langle n \rangle}, 1 \rangle$), a set $PRED_{\mathcal{L},n}$ of *n-ary predicates* (of type $\langle 1^{\langle n \rangle}, 0 \rangle$), and a set $OPER_{\mathcal{L},n}$ of *n-ary operators* (of type $\langle 1^{\langle n \rangle}, 1 \rangle$).

In particular, we may assume that whenever $\mathcal{L}$ is an elementary language, there exist a family of (possibly empty) sets $Q_{\mathcal{L}} = \{Q_{\mathcal{L},n} : n > 0\}$ and a denumerable subset $V_{\mathcal{L}} \subseteq NAME_{\mathcal{L}}$ such that, for each $n > 0$:

$$\{\langle x, y \rangle : \langle x, y \rangle \in CON_{\mathcal{L},n}\} = Q_{\mathcal{L},n} \times V_{\mathcal{L}}.$$

(Note that $Q_{\mathcal{L}}$ and $V_{\mathcal{L}}$ must be unique.) In that case, we may speak of $V_{\mathcal{L}}$ generically as a set of *name variables*, referring to an n-ary connective of the form $\langle q, v \rangle$ as an *n-ary quantifier binding* the variable v. In other words, we may treat quantifiers as a special kind of "structured" functors making sentences out of sentences, as long as we take one quantifier for each variable. (There are other ways of dealing with quantifiers within the present framework, but this one is particularly effective as it does not require any *ad hoc* account of the variable/constant dichotomy. The same would not be true, for instance, if we treated quantifiers as mixed functors of type $\langle 1, \langle 0^{\langle n \rangle}, 0 \rangle \rangle$.[10])

---

[10] See Levin [1982] for various other options. The present treatment is not without limitations. For instance, it does not allow one to single out elementary languages in which "vacuous" quantification does not arise (i.e., in which no quantifier can combine with a sentential expression in which the relevant variable does not occur "free"). However, this reflects a more general feature of the structural operation **g** and is not peculiar to variable binding: it is equally impossible to single out sentential languages where, say, "useless" conjunctions such as $A \wedge A$ are forbidden. A general solution would involve revising 1.1.2 by allowing **g** to be defined only for *some* arguments of the right types. To illustrate, define the set of *free variables* of an expression x, $\phi_{\mathcal{L}}(x)$, in the obvious way: (i) if $x \in SYM_{\mathcal{L}}$, then $\phi_{\mathcal{L}}(x) = \varnothing$; (ii) if $x = \mathbf{g}(y, z)$ and $y \notin \bigcup Q_{\mathcal{L}} \times V_{\mathcal{L}}$, then $\phi_{\mathcal{L}}(x) = \phi_{\mathcal{L}}(y) \cup \phi_{\mathcal{L}}(z)$; and (iii) if $x = \mathbf{g}(y, z)$ and $y = \langle q, v \rangle$ for some $q \in \bigcup Q_{\mathcal{L}}$ and some $v \in V_{\mathcal{L}}$, then $\phi_{\mathcal{L}}(x) = \phi_{\mathcal{L}}(z) - \{v\}$. Then one can rule out vacuous quantification by requiring $\mathbf{g}(\langle q, v \rangle, A)$ to be undefined if (and only if) $v \notin \phi_{\mathcal{L}}(A)$.

As before, we can then introduce some familiar notation by definition. Thus, suppose $\mathcal{L}$ is an elementary language whose symbols include a dyadic quantifier $\langle\downarrow,v\rangle$. Then an expression of the form

$$\mathbf{g}(\mathbf{g}(\langle\downarrow,v\rangle,A),B)$$

(where $A,B\in\mathbf{E}_0$) may naturally be written as

$$(A\downarrow^v B).$$

Indeed, suppose that the quantifiers $\{\langle\downarrow,v\rangle: v\in V_{\mathcal{L}}\}$ are the only connectives of $\mathcal{L}$. Then we may consider the following way of defining the other familiar connectives and quantifiers:

(a) '$(\neg A)$' for '$(A\downarrow^u A)$'
(b) '$(A\vee B)$' for '$(\neg(A\downarrow^w B))$'
⋮ ⋮
(f) '$(\bigvee vA)$' for '$(\neg(A\downarrow^v A))$'
(g) '$(\bigwedge vA)$' for '$((\neg A)\downarrow^v(\neg A))$'

where u is the first element of $V_{\mathcal{L}}-\omega_{\mathcal{L}}(A)$ (in the alphabetic order) and w the first element of $V_{\mathcal{L}}-(\omega_{\mathcal{L}}(A)\cup\omega_{\mathcal{L}}(B))$. If each $\langle\downarrow,v\rangle$ is interpreted as the *joint-denial* quantifier 'for anything v, neither . . . nor', the notation introduced in (a), (b), etc. corresponds to the connectives for *negation*, *disjunction*, etc., and the definitions in (f) and (g) correspond to the *particular* and the *universal* quantifier, respectively. We shall see below how this interpretation can be characterized semantically.[11] (As in 1.1.4 (A), this example is minimalistic: one may of course consider elementary languages that contain any number of connectives and quantifiers —typically one-place quantifiers such as the ones just mentioned—as genuine symbols.)

(C) $\mathcal{L}$ is a *full categorial language* if and only if $\mathcal{RD}\mathbf{s}=\mathcal{T}$, i.e., if and only if $SYM_{\mathcal{L}}$ comprises a non-empty set $S_{\mathcal{L},t}$ of symbols for each type $t\in\mathcal{T}$.

In particular, we may assume that if $\mathcal{L}$ is a full categorial language of this sort, then for each type $t\in\mathcal{T}$ there exist a family of (possibly

[11] See examples 1.2.4(B) and 1.3.4(B). Definitions (a)–(g) come from Schönfinkel [1924].

empty) sets $A_{\mathcal{L},t}$ disjoint from $SYM_{\mathcal{L}}$ and a denumerable set $V_{\mathcal{L},t} \subseteq S_{\mathcal{L},t}$ such that, for each type $t' \in \mathcal{T}$,

$$\{\langle x,y\rangle : \langle x,t',y\rangle \in S_{\mathcal{L},\langle t',\langle t,t'\rangle\rangle}\} = A_{\mathcal{L},t} \times V_{\mathcal{L},t}.$$

(Note that $A_{\mathcal{L},t}$ and $V_{\mathcal{L},t}$ must be unique for each $t \in \mathcal{T}$.) In that case we may speak of each $V_{\mathcal{L},t}$ as a set of *variables of type t*, referring to any structured symbol of the form $\langle a,t',v\rangle$ (where $\langle a,v\rangle \in A_{\mathcal{L},t} \times V_{\mathcal{L},t}$) as an *abstractor of type* $\langle t',\langle t,t'\rangle\rangle$ binding the variable v.

Again, with respect to such languages we can then introduce some familiar notation. For example, suppose $\mathcal{L}$ is a full categorial language with an abstractor $\langle \lambda,t',v\rangle$ for each variable v and each type $t' \in \mathcal{T}$ (where λ is some fixed object); then an expression of the form:

$$\mathbf{g}(\langle \lambda,t',v\rangle,A)$$

(where $A \in \mathbf{E}_{t'}$) may be written simply as

$$(\lambda vA).$$

With each abstractor $\langle \lambda,t',v\rangle$ interpreted as the *functional* abstractor of type $\langle t',\langle t,t'\rangle\rangle$, we may also introduce the following notation:

(a) '$\mathbf{I}_t$' for '$(\lambda v_\chi v_\chi)$'
(b) '$\mathbf{K}_{t,t'}$' for '$(\lambda v_\chi(\lambda v_{\chi'} v_\chi))$'
(c) '$\mathbf{S}_{t,t',t''}$' for '$(\lambda v_\alpha(\lambda v_\beta(\lambda v_\chi((v_\alpha(v_\chi))(v_\beta(v_\chi))))))$'

where $t,t',t'' \in \mathcal{T}$ and v_χ, $v_{\chi'}$, v_β, and v_α are fixed elements of $V_{\mathcal{L},t}$, $V_{\mathcal{L},t'}$, $V_{\mathcal{L},\langle t,t'\rangle}$, and $V_{\mathcal{L},\langle t,\langle t',t''\rangle\rangle}$, respectively. This notation parallels the usual characterization of the combinators for *identity*, *cancellation*, and *distribution*.[12]

1.1.5. AUXILIARY NOTIONS

I shall conclude this preliminary section with a couple of supplementary definitions that will prove useful in some later sections. Let $\mathcal{L}, \mathcal{L}' \in \mathbb{L}$. Then:

---

[12] The intended interpretation of the functional abstractor is discussed below, examples 1.2.4(C) and 1.3.4(C). On (a)–(c) see Sanchis [1964].

(a) $\mathcal{L}$ is *similar* to $\mathcal{L}'$ [written $\mathcal{L} \simeq \mathcal{L}'$] if and only if $\mathcal{D}\mathbf{s} = \mathcal{D}\mathbf{s}'$;

(b) $\mathcal{L}$ is a *reduction of* $\mathcal{L}'$, and $\mathcal{L}'$ an *expansion of* $\mathcal{L}$ [written $\mathcal{L} \preceq \mathcal{L}'$ and $\mathcal{L}' \succeq \mathcal{L}$, respectively], if and only if $\tau_{\mathcal{L}} \upharpoonright \mathcal{R}\mathbf{s} \subseteq \tau_{\mathcal{L}} \upharpoonright \mathcal{R}\mathbf{s}'$ and $\mathbf{g} \subseteq \mathbf{g}'$.

Note that both $\simeq$ and $\preceq$ (or $\succeq$) are pre-orderings of $\mathbb{L}$. More precisely, $\simeq$ is an equivalence relation whereas $\preceq$ is a partial ordering. And note that $\mathcal{L} \simeq \mathcal{L}'$ implies $\mathcal{D}\mathbf{E} = \mathcal{D}\mathbf{E}'$ while $\mathcal{L} \preceq \mathcal{L}'$ implies $\mathbf{E}_t \subseteq \mathbf{E}'_t$ for all $t \in \mathcal{D}\mathbf{E}$.

1.2. MODELS

Turning now to the notion of a model (for a given language), let me again remark that my approach is not meant to introduce any radical departure from the customary account. However, my aim is generality, so certain traditional constraints on how a language may be interpreted will be removed from the basic apparatus.

1.2.1. PRELIMINARIES

Let us first see what the customary account would look like. The guiding idea is that a model must act as a sort of semantic lexicon: it must determine what sort of things may be assigned to the basic components of the given language as their semantic counterparts, and it must do so in conformity to the relevant type distinctions. Thus, relative to our general definition of $\mathbb{L}$, a model for a language $\mathcal{L} = (\mathbf{s},\mathbf{g},\mathbf{E})$ would typically be a structure $\mathcal{M} = (\mathbf{d},\mathbf{h},\mathbf{I})$, with $\mathbf{d}$, $\mathbf{h}$, and $\mathbf{I}$ a triple of functions satisfying essentially the same conditions as $\mathbf{s}$, $\mathbf{g}$, and $\mathbf{E}$, respectively. That is, a typical model $\mathcal{M}$ for $\mathcal{L}$ would be characterized in such a way that (i) to each category of expressions $\mathbf{E}_t \in \mathcal{R}\mathbf{E}$ there corresponds some nonempty domain of interpretation $\mathbf{I}_t \in \mathcal{R}\mathbf{I}$; (ii) to each symbol $\mathbf{s}(\beta,t) \in \mathbf{E}_t$ there corresponds a denotation $\mathbf{d}(\beta,t) \in \mathbf{I}_t$; and (iii) to the structural operation $\mathbf{g} \subseteq (\bigcup \mathcal{R}\mathbf{E})^3$ there corresponds a structural operation $\mathbf{h} \subseteq (\bigcup \mathcal{R}\mathbf{I})^3$ subject to the same type restrictions, i.e., such that $x \in \mathbf{I}_{\langle t',t \rangle}$ and $y \in \mathbf{I}_{t'}$ always imply $\mathbf{h}(x,y) \in \mathbf{I}_t$. Such a characterization would be patterned directly after the language's structure. For the intuitive idea is that a model represents a conceivable state of affairs, a way things could be, and such a notion can only be represented to the extent that it can be decoded from the language itself.

In practice, of course, the intuitive appeal of the account would depend on the actual make up of the functions **d**, **h**, and **I**. For instance, it is customary to associate the type 0 with a domain of truth-values (typically the values *true* and *false*), the type 1 with a domain of individuals, and the corresponding derived types with higher-order entities of some sort such as sets or properties (type $\langle 1,0\rangle$), truth-functions (type $\langle 0,0\rangle$), operations (type $\langle 1,1\rangle$), and so on. Accordingly, if $\mathcal{D}$**s** contained indices of the form $\langle\gamma,0\rangle$, $\langle\zeta,1\rangle$, $\langle\xi,\langle 1,0\rangle\rangle$, etc., then $\mathbf{d}(\gamma,0)$ would be the truth-value in $\mathbf{I}_0$ associated with the sentence $\mathbf{s}(\gamma,0)$; $\mathbf{d}(\zeta,1)$ would be the individual in $\mathbf{I}_1$ designated by the name $\mathbf{s}(\zeta,1)$; $\mathbf{d}(\xi,\langle 1,0\rangle)$ would be the property in $\mathbf{I}_{\langle 1,0\rangle}$ corresponding to the predicate $\mathbf{s}(\xi,\langle 1,0\rangle)$; and so on. Clause (iii) above would then reflect the intuition that insofar as the result of applying a predicate to a name *via* **g** is always a sentence, likewise the result of ascribing a property to an individual via **h** should always be a truth-value—and similarly for the other cases.[13] For example, just as the result of applying the structural operation of English to the predicate 'cries' and the name 'Albert' yields the sentence 'Albert cries', the result of applying the corresponding semantic operation to the property of crying and the individual Albert yields a truth-value—typically subject to the schema:

h(crying, Albert)=*true* iff Albert cries.

Likewise, if k is the interpretation of the connective 'it is not the case that', then the result of applying **h** to a pair consisting of k and a truth-value x will always yield a truth-value—typically subject to the schema:

$\mathbf{h}(k, x)=\mathit{true}$ iff $x=\mathit{false}$.

This picture can be refined in many ways. Often, the semantic counterpart of a linguistic expression is not a fixed entity, but may de-

---

[13] The standard notion of model is of course rooted in the work of Frege [1892] and Tarski [1933]. Note, however, that various non-classical conceptions will also qualify as "standard" in this sense. For instance, it is not required that $\mathbf{I}_0$ contain only the two classical truth-values, so one may obtain a many-valued model by letting $\mathbf{I}_0$ include all the relevant truth-values.

pend on various contextual factors. Modal and tensed expressions, for example, may call for an account in terms of world coordinates or time indices. Accordingly, one might want to construe a domain $\mathbf{I}_t$ as comprising, not entities of the relevant sort (truth-values for t=0, individuals for t=1, etc.), but rather functions that pick out such entities relative to a suitable set C of contexts (worlds, times). Thus, if C is a set of possible worlds (with a suitable relation of accessibility) and V the set of truth-values (*true* and *false*), a connective such as 'it is necessarily the case that' could be interpreted as an item k such that, for every function (proposition, in the usual terminology) p: $C \rightarrow V$ and every world $c \in C$:

$\mathbf{h}(k,p)(c)=true$ iff $p(c')=true$ for every $c' \in C$ accessible from c.

Likewise, if C is a set of times (with a suitable relation of precedence), a connective such as 'it will always be the case that' could be interpreted as an item k such that, for every p: $C \rightarrow V$ and every time $c \in C$:

$\mathbf{h}(k,p)(c)=true$ iff $p(c')=true$ for every $c' \in C$ later than c.

Even quantifiers and other variable binders may be interpreted as context-dependent expressions of this kind. Insofar as they can be construed as functors (as seen in Section 1.1.4), their semantics can be given in terms of contextual models in which the relevant contexts are functions assigning a value to the bound variables. For instance, the universal quantifier 'for anything v' may be interpreted as an item k such that, for every p: $C \rightarrow V$ and every value assignment $c \in C$:

$\mathbf{h}(k,p)(c)=true$ iff $p(c')=true$ for every $c' \in C$ that coincides with c except possibly for the value assigned to v.

Of course, in some cases it may be necessary to combine these accounts and construe the relevant contexts as n-tuples of features (world coordinates, time indices, value assignments, etc., as the case may be). The generalization is obvious.

Let us, indeed, assume this general format as a paradigm of the standard account.[14] For every model $\mathcal{M}=(\mathbf{d},\mathbf{h},\mathbf{I})$, let us assume the ex-

---

[14] This general format stems from the work of Lewis [1970] and Montague [1970a, 1970b], though the idea of context-dependent semantic values originated with Carnap [1947].

istence of a set C of contexts and a system $\langle \mathbf{D}_t: t \in \mathcal{DI} \rangle$ of basic domains such that $\mathbf{I}_t = \mathbf{D}_t{}^C$ for each $t \in \mathcal{DI}$. (The simpler format can always be retrieved as the special case where C is a singleton.) This paradigm is definitely in line with the notion of a model that I want to consider. It is defined in general terms, regardless of the properties that a particular language or class of languages might enjoy. And it does not require any specific commitment concerning the nature of a model's domains. The framework as such does not depend on such philosophically controversial notions as 'truth-value', 'individual', 'property', etc., nor does it depend on the nature of the items to be included in the relevant package of contextual features. This is important if we aim at an account that is philosophically neutral. (For the same reason, in the definition of a language such basic concepts as 'sentence', 'name', 'predicate', etc. were not defined independently, but only in relation to the auxiliary notion of a type. Even the basic notions of 'symbol' and 'expression' are entirely language-relative.)

It is also clear, however, that such an account would incorporate the doubtful assumptions mentioned earlier in the Introduction and relative to which I wish, in the present context, to remain neutral. First of all, the requirement (i) that each domain of interpretation $\mathbf{I}_t$ be a *non-empty* set—in fact, a set of functions ranging over a non-empty basic domain $\mathbf{D}_t$—rules out the possibility of models in which certain categories of expressions fail to be instantiated. This is common practice; but it embeds a form of ontological commitment that seems to be extrinsic to pure semantic theorizing. For instance, with regard to expressions of type 1 (names) the non-emptiness requirement acquires a metaphysical connotation to the effect that there must be something rather than nothing.[15] Secondly, the requirement (ii) that each symbol $\mathbf{s}(\beta,t) \in \mathbf{E}_t$ denote an element of $\mathbf{I}_t$—i.e., a function yielding a unique value $\mathbf{d}(\beta,t)(c) \in \mathbf{D}_t$ for each $c \in C$—excludes the possibility that a symbol may, in some con-

[15] This was Wittgenstein's concern in [1913], a concern that eventually led to the development of so-called inclusive logics (see Church [1951], Mostowski [1951], Hailperin [1953], and Quine [1954] for the seminal papers). I am not aware of any similar work (or even similar concerns) regarding domains of type $t \neq 1$, though some authors have considered dispensing with the assumption that there exist possible worlds, hence that $\mathbf{I}_0 \neq \varnothing$ (see Lambert, Leblanc & Meyer [1969] and Morgan [1970]).

texts, remain without a semantic value, or receive more than one value. Again, this is common practice, but there are circumstances which seem to call for a looser account. A name *aims* to denote a unique individual in every context, but may fail to do so; a sentence aims to pick out a unique truth-value in every context, but may fail; and so on.[16] Likewise, the requirement (iii) that **g** be represented in the model by an operation **h** which assigns an element of $\mathbf{I}_t$ to each eligible pair of arguments $x \in \mathbf{I}_{\langle t',t\rangle}$ and $y \in \mathbf{I}_{t'}$ implies the existence of a unique value $\mathbf{h}(x,y)(c) \in \mathbf{D}_t$ for each context $c \in C$. This leaves no room for the various sorts of indeterminacy and overdeterminacy mentioned in the Introduction (such as those involved in the model stemming from the Holmes stories). Ideally, a model should indeed specify, for each relevant context, whether a given object possesses a given property; whether a certain object is the result of applying a given operation to another object; and so on. But a model may not—and need not—be so effective.[17] In this regard, the fact

---

[16] Such failures have been widely considered in the literature with regard to symbols of both types 0 and 1. On the one hand, the possibility of *bona fide* denotational "gaps" has been given serious thought since Strawson [1950] and has inspired various partial semantics allowing for non-denoting names and/or truth-valueless sentences (see Smiley [1960], van Fraassen [1966a], and Woodruff [1970] for seminal works, and Blamey [1986], Langholm [1988], and Fenstad [1997] for surveys and comparisons; the possibility of non-denoting names is also contemplated in Kripke's [1963] semantics for modal logics and has been the main rationale for the development of so-called free logics, i.e., logics free from existential presuppositions: see Leonard [1956], Leblanc & Hailperin [1959], Hintikka [1959], Lambert [1963], and Schock [1964] for pioneering works, and Garson [1984] and Bencivenga [1986] for overviews.) Some three-valued semantics (like those of Łukasiewicz [1920], Kleene [1938], or Bochvar [1939]) may also be viewed in this light, though the gap is actually treated as a third value. On the other hand, the possibility of denotational "gluts" has been taken seriously since Jaśkowski [1948] and has led to the development of various paraconsistent semantics (seminal works include the "antinomic" semantics of Asenjo [1966], Grant [1975], Da Costa & Alves [1976], and Priest [1979], or the semantics for relevant logic of Routley & Routley [1972] and Dunn [1976]; see Priest & Routley [1989a, 1989b] and Anderson, Belnap & Dunn [1992], respectively, for overviews). Work in the tradition of situation semantics (Barwise & Perry [1983], Barwise & Etchemendy [1987]) as well as developments in artificial intelligence (e.g., Fitting [1987], Patel-Schneider [1989], Léa Sombé [1990], Roos [1992], Fagin, Halpern, Moses & Vardi [1995]) have also been in the spirit of such generalizations.

[17] In this connection, discussion in the philosophical literature has generally focused on the case $\langle t',t\rangle = \langle 1^{\langle n\rangle},0\rangle$, i.e., where the first argument of **g** is a predicate. For

that the grammatical structure of any language in $\mathbb{L}$ is assumed to be fully determined—in that $\mathbf{g}(x,y)$ is always a well-defined expression when x and y are of the appropriate types—is of little import. Some expressions may be more awkward than others, but it would be a mistake to burden the notion of well-formedness with the task of disentangling "meaningful" from "meaningless" expressions. This is a semantic issue that must be accounted for precisely by allowing the semantic operation **h** to yield results that are not equally well-behaved.

There are, to be sure, various ways in which one could try to accommodate these desiderata within the classical framework outlined above. As I mentioned, however, the approach that I wish to consider here is more explorative: I want to take a look at the picture that results from giving up these assumptions altogether (regardless of any specific applications). Thus, I will first of all allow the basic domains of a model to be arbitrary, possibly empty sets. Second, instead of a total function f: $C \to \mathbf{D}_t$, I shall simply assume the denotation of a symbol $\mathbf{s}(\beta,t) \in \mathbf{E}_t$ to be a relation $R \subseteq C \times \mathbf{D}_t$ associating each context $c \in C$ with zero, one, or more values in $\mathbf{D}_t$. We may think of such a relation as an "approximate function": it approximates membership in $\mathbf{I}_t$, but it may fail to do so insofar as it may be undefined or overdefined in some contexts. Likewise, rather than a total function f: $C \to \mathbf{D}_t$, I shall allow the result of applying **h** to an eligible pair of arguments $x \in \mathbf{I}_{\langle t',t \rangle}$ and $y \in \mathbf{I}_{t'}$ to be an arbitrary relation $R \subseteq C \times \mathbf{D}_t$—a relation approximating membership in $\mathbf{I}_t$.[18]

---

instance, various accounts of vagueness and sortal incorrectness make use of models in which certain predicates are allowed to be "indeterminate" with respect to some objects in the domain, corresponding to the possibility that the value of **h** for some of its arguments be a *partial* function (see Thomason [1972], Fine [1975], Przełęcki [1976], Kamp [1975, 1981a], Klein [1980], and Pinkal [1983] for illustrative examples, and Keefe & Smith [1997] for an overview). This feature is shared by theories dealing with a partial truth predicate in the spirit of Kripke [1975b] and Martin & Woodruff [1975]. On the other hand, there are theories in which the truth predicate is allowed to have a glut rather than a gap (e.g., Dunn [1976], Priest [1987, 1989], Visser [1984]), and some authors have considered treating vague predicates in a similar way (e.g., Arruda & Alves [1979], Peña [1989], Hyde [1997]). Such accounts correspond to the possibility that the value of **h** for some arguments be not a function but a proper *relation*.

[18] The idea of thinking of relations as approximate functions is due to Scott [1972] (compare Feferman [1984], Woodruff [1984b], and Fitting [1989]). Notice

Generally speaking, this way of generalizing the standard account corresponds to the idea that in some models the values of **d** and **h** may fail to be proper elements of $\bigcup\mathcal{R}\mathbf{I}$. One could, in fact, construe $\bigcup\mathcal{R}\mathbf{I}$ itself as a set of relations, taking $\mathbf{I}_t \subseteq \wp(C\times\mathbf{D}_t)$ for each $t\in\mathcal{D}\mathbf{I}$. That would give us something mathematically closer to standard models, but it strikes me as a misleading way of proceeding. It would imply, for instance, that in order to allow for the possibility that a sentence symbol lack a unique truth-value in a context c (i.e., the possibility that $\mathbf{d}(\beta,t)[c]$ contain no truth-value or more than one truth-value, where $\mathbf{s}(\beta,t)$ is the symbol in question) we must *ipso facto* provide for a different interpretation of the connectives; for these would immediately become operations on relations rather than on functions (propositions). This is unjustified. We may interpret connectives that way, but we need not do so. From a different point of view, construing the domains of interpretation as sets of relations between contexts and semantic values would amount to the idea that models which are somewhat incomplete or inconsistent are "ontologically" so. Such models would involve incomplete/inconsistent propositions, incomplete/inconsistent individual functions, etc. By contrast, I take the issue to be purely semantic. A sentence *aims* to pick out a proposition—to have exactly one truth-value in every context—but it may fail; and it may fail to a greater or lesser degree depending on the extent to which its denotation behaves like a true proposition in $\mathbf{I}_0$. Likewise, a name aims to denote a unique individual in every context, but it may fail; and it may fail to the degree to which its denotation approximates an element of $\mathbf{I}_1$. And so on.[19]

---

that our decision to treat n+2-adic functors as a kind of monadic functors might in this regard give rise to some oddities. For instance, let $x_1,x_2\in\mathbf{I}_{\langle 1,\langle 1,1\rangle\rangle}$ and $y\in\mathbf{I}_1$ and suppose that (i) $\mathbf{h}(x_1,y)=\varnothing$ while $\mathbf{h}(x_2,y)=v$ for some $v\in\mathbf{I}_{\langle 1,1\rangle}$ so that $\mathbf{h}(v,w)=\varnothing$ for all $w\in\mathbf{I}_1$; and (ii) for all $z\in\mathbf{I}_1-\{y\}$, $\mathbf{h}(x_1,z)=\mathbf{h}(x_2,z)=v$ for some $v\in\mathbf{I}_1$ so that $\mathbf{h}(v,w)=w$ for all $w\in\mathbf{I}_1$. Then x_1 can be regarded as coding the interpretation of a dyadic functor F whose value for all pairs $\langle z,w\rangle\in\mathbf{I}_1\times\mathbf{I}_1$ is w if $z\neq y$, and undefined otherwise. But if x_2 is to code anything at all, it must code the interpretation of F too. Yet $x_1\neq x_2$. (The example is adapted from Tichy [1982]; see also Muskens [1995], ch. 2.) Luckily this ambiguity has no effects on what follows, so there is no need to complicate things.

[19] This is not to say that the approach that I am dismissing (effectively, the one followed by the authors mentioned in note 18) is incompatible with the present account. To the contrary, that approach corresponds to a special case where all models are, in a sense, standard. I shall come back to this in section 1.2.3 below.

To sum up, then, let us agree to call a sequence **I** a *system of contextual sets* if and only if there exist a set $C \neq \varnothing$ and a family of sets $\mathbf{D} = \langle \mathbf{D}_j : j \in \mathcal{D}\mathbf{I} \rangle$ such that

$$\mathbf{I}_j = \mathbf{D}_j^C \text{ for all } j \in \mathcal{D}\mathbf{I}.$$

(I shall call C the *context set* for **I** and each $\mathbf{D}_j$ the *basis* of the corresponding $\mathbf{I}_j$.) And let us say that a relation R *approximates membership* in a contextual set $\mathbf{D}_j^C$ (written $R \underset{\sim}{\in} \mathbf{D}_j^C$) if $\mathcal{D}R \subseteq C$ and $\mathcal{R}R \subseteq \mathbf{D}_j$. Then the kind of general structures that I am going to assume as models for our languages may be defined as follows.

1.2.2. DEFINITION

A *model* for a language $\mathcal{L} = (\mathbf{s}, \mathbf{g}, \mathbf{E})$ is a structure $\mathcal{M} = (\mathbf{d}, \mathbf{h}, \mathbf{I})$ satisfying the following general conditions:

(a) **d** is a function with $\mathcal{D}\mathbf{d} = \mathcal{D}\mathbf{s}$;
(b) **h** is also a function;
(c) **I** is a system of contextual sets closed under the properties: (i) if $\langle \beta, t \rangle \in \mathcal{D}\mathbf{s}$, then $t \in \mathcal{D}\mathbf{I}$ and $\mathbf{d}(\beta, t) \underset{\sim}{\in} \mathbf{I}_t$; and (ii) if $t, t' \in \mathcal{D}\mathbf{I}$, $x \in \mathbf{I}_t$, $y \in \mathbf{I}_{t'}$, and $t = \langle t', t'' \rangle$, then $t'' \in \mathcal{D}\mathbf{I}$ and $\mathbf{h}(x, y) \underset{\sim}{\in} \mathbf{I}_{t''}$.

(For definiteness, I shall always assume that **h** is smallest among the relations that satisfy (c), so that $\mathbf{h}(x,y)$ is not defined unless x and y comply with the stated conditions.)

The relevant terminology should be clear from the foregoing discussion. If $\mathcal{M} = (\mathbf{d}, \mathbf{h}, \mathbf{I})$ is a model for a language $\mathcal{L} = (\mathbf{s}, \mathbf{g}, \mathbf{E})$, the elements of $\mathcal{R}\mathbf{d}$ are the *denotation assignments* (in $\mathcal{M}$) associated with the symbols in $\mathcal{R}\mathbf{s}$, **h** is the *structural operation* corresponding (in $\mathcal{M}$) to the operation **g**, and $\bigcup \mathcal{R}\mathbf{I}$ serves as a domain of *interpretation* (in $\mathcal{M}$) for the expressions in $\bigcup \mathcal{R}\mathbf{E}$. In particular, I shall take the basis $\mathbf{D}_t$ of each domain $\mathbf{I}_t$ ($t \in \mathcal{D}\mathbf{I}$) as comprising the possible semantic values for the expressions in the corresponding category $\mathbf{E}_t$ (note that $\mathcal{D}\mathbf{I} = \mathcal{D}\mathbf{E}$). If c is any member of the context set C for **I**, I shall then speak of the values in $\mathbf{d}(\beta, t)[c]$ (for $\langle \beta, t \rangle \in \mathcal{D}\mathbf{s}$) as the *denotations of* $\mathbf{s}(\beta, t)$ in context c, and of the elements of $\mathbf{h}(x, y)[c]$ (for x and y as appropriate) as the *values* obtained by applying x to y in context c.

Note that, in general, the conception of C as a set of contextual features suggests that $\mathbf{h}(x,y)[c]$ may also be thought of as comprising the values of applying $x(c)$ to $y(c)$ (in some intuitive sense). For instance, if x is of type $\langle 1,0\rangle$ and y of type 1, then to say that $\mathbf{h}(x,y)[c]$ contains (at least) the value *true* is to say that the model represents the property $x(c)$ as being (at least) true *of* the individual $y(c)$. Likewise, if x is of type $\langle 1,1\rangle$ and y of type 1, then to say that $\mathbf{h}(x,y)[c]$ contains (at least) the value z is to say that the model represents $x(c)$ approximately as an operation which, when applied to the individual $y(c)$, yields (at least) the output z. This means that, in general, it is natural to suppose that $\mathbf{h}$ satisfies two supplementary conditions. First, the outputs of $\mathbf{h}$ should not depend on C unless its arguments do. That is, given $x,y \in \bigcup\mathcal{R}\mathbf{I}$ and $c,c' \in C$, it is natural to suppose that

if $x(c)=x(c')$ and $y(c)=y(c')$, then $\mathbf{h}(x,y)[c]=\mathbf{h}(x,y)[c']$.

Second, the outputs of $\mathbf{h}$ should always behave coherently with respect to the functions that make up the domains of interpretation. Thus, given $x,x',y,y' \in \bigcup\mathcal{R}\mathbf{I}$ and $c \in C$, the following conditionals should hold too:

if $x(c)=x'(c)$, then $\mathbf{h}(x,y)[c]=\mathbf{h}(x',y)[c]$
if $y(c)=y'(c)$, then $\mathbf{h}(x,y)[c]=\mathbf{h}(x,y')[c]$.

In the following, all models will be assumed to satisfy these supplementary conditions unless otherwise specified. And I shall speak of the elements of $\mathbf{h}(x,y)[c]$ indifferently as the values of applying x to y in c or as the values of applying $x(c)$ to $y(c)$ (according to the given model).

1.2.3. REMARKS

Given a language $\mathcal{L}$ and a set C, the class of all models for $\mathcal{L}$ with context set C will be denoted by '$MOD_{\mathcal{L},C}$'. Thus, $MOD_{\mathcal{L}}=\bigcup\{MOD_{\mathcal{L},C}: C\neq\varnothing\}$ will be the class of all models for $\mathcal{L}$, and $\mathbb{M}=\bigcup\{MOD_{\mathcal{L}}: \mathcal{L}\in\mathbb{L}\}$ the class of all models. I shall employ the letter '$\mathcal{M}$', possibly with subscripts or superscripts, exclusively as a variable on $\mathbb{M}$, with the understanding that $\mathcal{M}=(\mathbf{d},\mathbf{h},\mathbf{I})$, $\mathcal{M}'=(\mathbf{d}',\mathbf{h}',\mathbf{I}')$, $\mathcal{M}_j=(\mathbf{d}_j,\mathbf{h}_j,\mathbf{I}_j)$, and so on. These conventions are symmetric to those of Section 1.1.3 and share with them some innocuous notational ambiguity (due to the fact that '$\mathbf{I}_j$' may denote the third component of a model $\mathcal{M}_j$ as well as a basic domain of a model $\mathcal{M}$: the context will always prevent confusion). Similar conventions will apply to the structures associated with the domains of interpretation of each model, so that C, C', and C_j will be the context sets for $\mathbf{I}$, $\mathbf{I}'$, and $\mathbf{I}_j$, respectively, and $\mathbf{D}$, $\mathbf{D}'$, and $\mathbf{D}_j$ the corresponding systems of basic domains. (When necessary, the context set of a model $\mathcal{M}$ will be denoted more explicitly by '$C_{\mathcal{M}}$'.) Note also that $\mathbb{L}\cap\mathbb{M}\neq\varnothing$, i.e., some languages qualify as models for themselves and some models qualify as languages. However, nothing incongruous will arise from this fact.

Some further notation and terminology will be useful. First of all, as a result of our generalization the models of any language $\mathcal{L}$ may naturally be classified into two major (overlapping) classes. On the one hand, some models provide the language with a complete interpretation: they assign *at least one* denotation per context to each symbol, and their structural operation is such as to assign *at least one* value per context to each possible pair of arguments of the appropriate type. More precisely, the class of *complete* models for a given language $\mathcal{L}$ (written $\mathit{COMP}_{\mathcal{L}}$) can be singled out as comprising those elements $\mathcal{M}\in \mathit{MOD}_{\mathcal{L}}$ for which the following requirements hold jointly:

for all $\langle\beta,t\rangle\in\mathcal{D}\mathbf{s}$ and all $c\in C_{\mathcal{M}}$: $\mathbf{d}(\beta,t)[c]\gtrsim 1$
for all $\langle x,y\rangle\in\mathcal{D}\mathbf{h}$ and all $c\in C_{\mathcal{M}}$: $\mathbf{h}(x,y)[c]\gtrsim 1$.

(Of course, a necessary condition for these requirements to hold is that $\mathbf{D}_t\neq\varnothing$ for each $t\in\mathcal{D}\mathbf{E}$.) By contrast, there are models which provide the language with a consistent interpretation throughout: they assign *at most one* denotation per context to each symbol, and their structural operation assigns *at most one* value per context to each eligible pair of arguments. In other words, to each language $\mathcal{L}$ there corresponds a class of *consistent* models (written $\mathit{CONS}_{\mathcal{L}}$) which comprises exactly those elements $\mathcal{M}\in \mathit{MOD}_{\mathcal{L}}$ such that

for all $\langle\beta,t\rangle\in\mathcal{D}\mathbf{s}$ and all $c\in C_{\mathcal{M}}$: $\mathbf{d}(\beta,t)[c]\lesssim 1$
for all $\langle x,y\rangle\in\mathcal{D}\mathbf{h}$ and all $c\in C_{\mathcal{M}}$: $\mathbf{h}(x,y)[c]\lesssim 1$.

(Here, a sufficient though trivial condition for these requirements to be satisfied is that $\mathbf{D}_t=\varnothing$ for each $t\in\mathcal{D}\mathbf{E}$.)

There is an obvious sense in which these two classes are dual to each other: they reflect the two major restrictions on the notion of a model that have been dismissed in the definition—namely, *totality* on the one hand, and *functionality* on the other. Thus, if we restrict our attention to those models that are both complete and consistent, we get exactly the *sharp* models of the standard account. In the following, the class of such models (relative to a given language $\mathcal{L}$) will be denoted by '$SHRP_{\mathcal{L}}$', and the fact that

$$SHRP_{\mathcal{L}} = COMP_{\mathcal{L}} \cap CONS_{\mathcal{L}} \subseteq MOD_{\mathcal{L}}$$

indicates one important sense in which the present account purports to subsume the standard one as a distinguished special case. (Our sharp models are in fact more general than "classical" models as usually understood, or as they would be defined relative to $\mathbb{L}$, but that is unessential: the point is simply that all classical models are sharp, and all sharp models are in $\mathbb{M}$.) The notation '$COMP_{\mathcal{L},C}$', '$CONS_{\mathcal{L},C}$', and '$SHRP_{\mathcal{L},C}$' will be used in a similar fashion to denote the relevant classes of complete, consistent, and sharp models, respectively, relative to a given set C of contextual features.

In effect, the generality of definition 1.2.2 lies precisely in the fact that any one of a variety of possible approaches to the semantics of a given language $\mathcal{L} \in \mathbb{L}$ translates directly into the choice of some particular subclass of $MOD_{\mathcal{L}}$. I have already mentioned, for example, that one could construe the domains of interpretation of a model as consisting of relations rather than functions, taking

$$\mathbf{I}_t \subseteq \wp(C \times \mathbf{D}_t)$$

for each $t \in \mathcal{D}\mathbf{I}$. I argued that this is not a good way of generalizing the standard account, as we should rather speak of relations *approximating* membership in $\bigcup \mathcal{R}\mathbf{I}$. However, this is not to say that such a course is incompatible with the present account. To the contrary, it corresponds to a simple special case. It corresponds to the case obtained by selecting only models that qualify as standard in the present sense (i.e., sharp) except that each basic domain $\mathbf{D}_t$ consists, not of entities of the relevant sort (truth-values for t=0, individuals for t=1, etc.), but of arbitrary sets thereof. Since I have not included this requirement in 1.2.2, the class of models satisfying it (call them *relational* models) does not coincide with the whole $MOD_{\mathcal{L}}$; yet it clearly constitutes a specifiable subclass of $MOD_{\mathcal{L}}$ —indeed a subclass of $SHRP_{\mathcal{L}}$.

Here is a similar example. In standard semantics it is common to think of a basic domain associated with a functional type $\langle t',t\rangle$ as a set of functions mapping the domain $\mathbf{D}_{t'}$ associated with the first coordinate of the type into the domain $\mathbf{D}_t$ associated with the second coordinate. Accordingly, we could generalize this conception by construing each domain of the form $\mathbf{D}_{\langle t',t\rangle}$ as a set of *relations*:

$$\mathbf{D}_{\langle t',t\rangle} \subseteq \wp(\mathbf{D}_{t'} \times \mathbf{D}_t).$$

In that case, the conditions on **h** become more specific: where $x \in \mathbf{I}_{\langle t',t\rangle}$, $y \in \mathbf{I}_{t'}$, and $c \in C_{\mathcal{M}}$, **h** will satisfy the equation

$$\mathbf{h}(x,y)[c]=x(c)[y(c)].$$

Again, this requirement is not part of definition 1.2.2; yet the models fulfilling it (call them *stratified* models) constitute a specifiable subclass of $MOD_{\mathcal{L}}$.

Similar considerations apply to the basic domains associated with the individual types $t \in \omega$. Definition 1.2.2 clearly allows these to be arbitrary sets, hence to vary from model to model. In practice, one is often interested only in the study of models which agree in assigning the same, fixed basic domain to certain basic categories—say, only models in which $\mathbf{D}_0$ (if defined) is a fixed set of truth-values. No such requirement is included in 1.2.2; but we may single out the relevant subclass of $MOD_{\mathcal{L}}$ by specifying the desired conditions. The same criterion may be used also to account for the fact that some symbols of the language (e.g., the "logical" symbols) are often assumed to have a fixed meaning, i.e., a denotation assignment subject to a fixed set of conditions: in general, such assumptions can be regarded as determining a selection of elements of $MOD_{\mathcal{L}}$ as the only admissible models, and different assumptions will correspond to different selections.[20]

I shall return to these general issues in the next chapter, when the overall picture will be more complete. For the moment, let me conclude by stressing that the general notion of a model defined in 1.2.2 includes as a special case those models that do not involve any kind of context-dependence. Such *extensional* models—as we may call them—may be

---

[20] For more discussion on this point, see Varzi [1994b].

identified with those models $\mathcal{M}$ whose context set C is a singleton. In this case the notation introduced above is slightly redundant. For instance, it would be more natural to speak of a symbol as having no denotations or more denotations in $\mathcal{M}$, rather than in the unique context of $\mathcal{M}$. However, no confusion should arise from this redundancy.

1.2.4. EXAMPLES

Let us briefly illustrate the foregoing with reference to the examples introduced in 1.1.4. Let $\mathcal{L}$ be any language, $\mathcal{M}$ a model for $\mathcal{L}$, and $k \geq 2$ a finite ordinal (we shall think of k as a set of truth-values, with $\subseteq$ ordering k in terms of increasing truthfulness). Then:

(A) If $\mathcal{L}$ is a sentential language, then $\mathcal{M}$ is a *k-valued sentential model* for $\mathcal{L}$ if and only if $\mathbf{D}_0 = k$.

In particular, suppose $FUN_{\mathcal{L}}$ contains only a binary connective $\downarrow$, as in Example 1.1.4(A); then we may say that $\mathcal{M}$ is *extensionally adequate* to $\mathcal{L}$ if and only if the following condition is well-defined and holds for every $x,y \in \mathbf{I}_0$ and every $c \in C_{\mathcal{M}}$:[21]

(*) if $\mathbf{s}(\beta,t) = \downarrow$, then $\mathbf{h}(\mathbf{h}(\mathbf{d}(\beta,t),x),y)(c) = (k-1)-(x(c) \cup y(c))$,

i.e., if and only if $\mathcal{M}$ interprets $\downarrow$ as the *joint denial* connective 'neither ... nor' relative to the given set of truth-values. (This is the interpretation that justifies the relative definitions of '$\neg$', '$\vee$', '$\wedge$', '$\rightarrow$', and '$\leftrightarrow$' suggested in 1.1.4(A). It is most intuitive when $\mathcal{M}$ is a 2-valued sharp model, in which case (*) amounts to the requirement that $\downarrow$ be interpreted as in classical 2-valued logic: it yields a proposition that is true (i.e., takes the value 1) just in those contexts where its arguments are both false (i.e., take the value 0).[22])

---

[21] Since $\mathcal{M}$ may not be sharp, $\mathbf{h}(\mathbf{h}(\mathbf{d}(\beta,t),x),y)(c)$ will be well-defined only if $\mathbf{d}(\beta,t) \in \mathbf{I}_{\langle 0,\langle 0,0\rangle\rangle}$, $\mathbf{h}(\mathbf{d}(\beta,t),x) \in \mathbf{I}_{\langle 0,0\rangle}$, and $\mathbf{h}(\mathbf{h}(\mathbf{d}(\beta,t),x),y) \in \mathbf{I}_0$; similar considerations apply to examples (B) and (C) below.

[22] See above, note 9. For $k=3$, (*) amounts to a corresponding requirement in the 3-valued logic of Kleene [1938]. More generally, for $k \geq 2$, (*) reflects a natural way of interpreting '$\downarrow$' as 'neither ... nor' in the k-valued logic of the sequence of Dienes [1949]. See also below, example 1.3.4(A).

(B) If $\mathcal{L}$ is an elementary language with a suitable set $V_{\mathcal{L}}$ of name variables, then $\mathcal{M}$ is a *k-valued elementary model* for $\mathcal{L}$ if and only if $\mathbf{D}_0=k$ and $C_{\mathcal{M}}=\mathbf{D}_1{}^{V_{\mathcal{L}}}$; if $\mathbf{D}_1\neq\varnothing$, then $\mathcal{M}$ is also said to be *non-empty*.

Thus, to account for the fact that some $\mathcal{L}$-expressions may involve *variable* symbols of type 1 we may require $\mathcal{M}$ to be a model with $\mathbf{D}_1{}^{V_{\mathcal{L}}}$ as a context set, i.e., we may take the domains of interpretation to consist of functions defined on the value assignments c: $V_{\mathcal{L}}\rightarrow\mathbf{D}_1$. As mentioned in Section 1.2.1, this will account for the intuition that the interpretation of such expressions may depend on the values of the variables occurring therein.[23] (More generally, we could take $C_{\mathcal{M}}=\mathbf{D}_1{}^{V_{\mathcal{L}}}\times A$, where A is a set of further (tuples of) relevant contextual features; the generalization is obvious and I shall ignore it.) In particular, suppose $FUN_{\mathcal{L}}$ includes a binary quantifier $\langle\downarrow,v\rangle$ for each variable $v\in V_{\mathcal{L}}$, as in Example 1.1.4(B); then we may say that $\mathcal{M}$ is *logically adequate* to $\mathcal{L}$ if and only if the following conditions are well-defined and hold for all $\langle\beta,t\rangle\in\mathcal{D}\mathbf{s}$, all $v\in V_{\mathcal{L}}$, all $x,y\in\mathbf{I}_0$, and all $c\in C_{\mathcal{M}}$:

(a) if $\mathbf{s}(\beta,t)\notin V_{\mathcal{L}}$, then $\mathbf{d}(\beta,t)[c']=\mathbf{d}(\beta,t)[c]$ for all $c'\in C_{\mathcal{M}}$
(b) if $\mathbf{s}(\beta,t)=v$, then $\mathbf{d}(\beta,t)(c)=c(v)$
(c) if $\mathbf{s}(\beta,t)=\langle\downarrow,v\rangle$, then
$$\mathbf{h}(\mathbf{h}(\mathbf{d}(\beta,t),x),y)(c)=\bigcap\{(k-1)-(x(c(v/u))\cup y(c(v/u))):\ u\in\mathbf{D}_1\},$$

i.e., if and only if $\mathcal{M}$ interprets each element of $V_{\mathcal{L}}$ as an individual variable and each symbol $\langle\downarrow,v\rangle$ as the *joint-denial quantifier* 'for anything v, neither ... nor', all other constant symbols having constant denotations of the right type. (Again, this is the interpretation that justifies the relative definitions of '$\neg$', '$\vee$', ..., '$\bigvee v$', and '$\bigwedge v$' given in 1.1.4(B). It is most intuitive when $\mathcal{M}$ is a 2-valued sharp model, in which case condition (c) amounts to the requirement that each quantifier $\langle\downarrow, v\rangle$ be interpreted as in classical 2-valued elementary logic: its denotation on an assignment c: $V_{\mathcal{L}}\rightarrow\mathbf{D}_1$ behaves as a binary operation that yields the value 1

[23] This way of exploiting the intensional character of quantifiers and other variable binders within a categorial setting was first pointed out in Lewis [1970] (though Lewis [1983a] marks a change of view in the spirit of Cresswell [1977], recommending that variable-binding be treated outside the scope of functional application). I have defended it in Varzi [1994a]. In general terms, the idea is rooted in the theory of cylindric algebras originated with Henkin [1950] and Tarski [1952].

just in case both of its arguments yield the value 0 on every v-variant of c, i.e., on every assignment $c' \in C_{\mathcal{M}}$ that coincides with c except possibly for the value assigned to v.[24])

(C) If $\mathcal{L}$ is a full categorial language with a set $V_{\mathcal{L},t}$ of variables for each type $t \in \mathcal{T}$, then $\mathcal{M}$ is a *(full) k-valued categorial model* for $\mathcal{L}$ if and only if $\mathbf{D}_0 = k$ and $C_{\mathcal{M}} = \prod\langle \mathbf{D}_t^{V_{\mathcal{L},t}}: t \in \mathcal{T}\rangle$.

Again, since an expression of $\mathcal{L}$ may include *variable* symbols of various types, we require here that $\mathcal{M}$ involve the appropriate set of contextual features—in this case, sequences of assignments c_t: $V_{\mathcal{L},t} \rightarrow \mathbf{D}_t$ for each $t \in \mathcal{T}$. In particular, let $V_{\mathcal{L}} = \bigcup\{V_{\mathcal{L},t}: t \in \mathcal{T}\}$ and suppose that $SYM_{\mathcal{L}}$ includes an abstractor $\langle\lambda, t', v\rangle$ for every variable $v \in V_{\mathcal{L}}$ and every type $t' \in \mathcal{T}$, as in Example 1.1.4(C); then we may say that $\mathcal{M}$ is *combinatorially adequate* to $\mathcal{L}$ if and only if the following conditions are well-defined and hold for all $\langle\beta,t\rangle \in \mathcal{D}$s, all $t' \in \mathcal{T}$, all $v \in V_{\mathcal{L}}$, all $x \in \mathbf{I}_{t'}$, all $y \in \mathbf{I}_{\tau(v)}$, and all $c \in C_{\mathcal{M}}$:

(a) if $\mathbf{s}(\beta,t) \notin V_{\mathcal{L}}$, then $\mathbf{d}(\beta,t)[c'] = \mathbf{d}(\beta,t)[c]$ for all $c' \in C_{\mathcal{M}}$
(b) if $\mathbf{s}(\beta,t) = v$, then $\mathbf{d}(\beta,t)(c) = c_{\tau(v)}(v)$
(c) if $\mathbf{s}(\beta,t) = \langle\lambda,t',v\rangle$, then $\mathbf{h}(\mathbf{h}(\mathbf{d}(\beta,t),x),y)(c) = x(c(\tau(v)/c_{\tau(v)}(v/y(c))))$,

i.e., if and only if $\mathcal{M}$ interprets each $\langle\lambda,t',v\rangle$ as the *functional abstractor* of type $\langle t',\langle t,t'\rangle\rangle$ binding the corresponding *variable* v, every *constant* symbol having constant denotations of the right sort. (Relative to sharp models, this interpretation reflects the semantics of functional abstraction in the standard typed lambda calculus, which provides intuitive grounds for the definitions of 'I_t', '$\mathsf{K}_{t,t'}$' and '$\mathsf{S}_{t,t',t''}$' given in example 1.1.4(C).[25])

1.2.5. AUXILIARY NOTIONS

I conclude this section by defining a couple of simple notions that will play a pivotal role in the remainder of this work. Let $\mathcal{M}, \mathcal{M}' \in \mathbb{M}$. Then:

---

[24] See above, note 11. For $k>2$, condition (c) provides a natural generalization of this interpretation to k-valued logics. See also below, example 1.3.4(B).

[25] See above, note 12, and below, example 1.3.4(C).

(a) $\mathcal{M}$ is *commensurate* to $\mathcal{M}'$ [written $\mathcal{M} \asymp \mathcal{M}'$] if and only if $\mathcal{D}\mathbf{d} = \mathcal{D}\mathbf{d}'$ and $C_{\mathcal{M}} = C_{\mathcal{M}'}$;

(b) $\mathcal{M}$ is a *restriction of* $\mathcal{M}'$, and $\mathcal{M}'$ an *extension of* $\mathcal{M}$ [written $\mathcal{M} \sqsubseteq \mathcal{M}'$ and $\mathcal{M}' \sqsupseteq \mathcal{M}$, respectively], if and only if (i) $\mathbf{d}(\beta,t) \subseteq \mathbf{d}'(\beta,t)$ for all $\langle\beta,t\rangle \in \mathcal{D}\mathbf{d}$, (ii) $\mathbf{h}(x,y) \subseteq \mathbf{h}'(x,y)$ for all $\langle x,y\rangle \in \mathcal{D}\mathbf{h}$, and (iii) $\mathbf{I}_t \subseteq \mathbf{I}_t'$ for all $t \in \mathcal{D}\mathbf{I}$.

Evidently $\asymp$ is an equivalence relation (two models are commensurate if and only if they belong to the same class $MOD_{\mathcal{L},C}$) while $\sqsubseteq$ is a partial ordering (due to the reflexivity, antisymmetry, and transitivity of $\subseteq$). Note also that clause (b)(iii) implies that the context set of a model is the same as the context set of its restrictions and extensions. Intuitively, we may thus think of $\sqsubseteq$ as an ordering going up-hill in terms of degrees of definiteness: if $\mathcal{M} \sqsubseteq \mathcal{M}'$, then $\mathcal{M}'$ will involve fewer gaps than $\mathcal{M}$ (if any) and $\mathcal{M}$ will involve fewer gluts than $\mathcal{M}'$ (if any).[26]

It is also worth observing that $\sqsubseteq$ is a special case of the more general relation of embeddability, defined thus:

(c) $\mathcal{M}$ is *embeddable in* $\mathcal{M}'$ [written $\mathcal{M} \precsim \mathcal{M}'$, or $\mathcal{M}' \sqsupseteq \mathcal{M}$] if and only if there is a function f such that (i) $f(\mathbf{d}(\beta,t)) \subseteq \mathbf{d}'(\beta,t)$ for all $\langle\beta,t\rangle \in \mathcal{D}\mathbf{d}$, (ii) $f(\mathbf{h}(x,y)) \subseteq \mathbf{h}'(f(x),f(y))$ for all $\langle x,y\rangle \in \mathcal{D}\mathbf{h}$, and (iii) $\{f(x) \colon x \in \mathbf{I}_t\} \subseteq \mathbf{I}_t'$ for all $t \in \mathcal{D}\mathbf{I}$.

A smallest function f satisfying conditions (i)–(iii) is called a *homomorphism* of $\mathcal{M}$ into $\mathcal{M}'$, and inspection shows that the relation $\mathcal{M} \precsim \mathcal{M}'$ holds if and only if the identity map $= \restriction (\mathcal{R}\mathbf{d} \cup \mathcal{R}\mathbf{h} \cup \bigcup \mathcal{R}\mathbf{I})$ is a homomor-

---

[26] It might be argued that $\sqsubseteq$ is not the most natural candidate for this intuitive reading. In the case of stratified models, for instance, one might find it more natural to require that if $\mathbf{s}(\beta,\langle t',t\rangle)$ is a functor, then a necessary condition for $\mathcal{M}'$ to be "more definite" than $\mathcal{M}$ is that the relation(s) picked out by $\mathbf{d}(\beta,\langle t',t\rangle)$ in a given context $c \in C$ be included in the relation(s) picked out in that context by $\mathbf{d}'(\beta,\langle t,t'\rangle)$, i.e., that $\mathbf{d}(\beta,\langle t',t\rangle)[c] \subseteq \{R \restriction \mathbf{D}_{t'} \colon R \in \mathbf{d}'(\beta,\langle t,t'\rangle)[c]\}$. However, there is an obvious correspondence between extensions (or restrictions) defined along these lines and extensions (or restrictions) as defined in 1.2.5. (See note 31 below.) There is also an intuitive and significant connection between the present notions of extension and restriction and the notions of expansion and contraction employed in the literature on belief revision (see Gärdenfors [1988] and Levi [1980, 1991] for the basic material, Fuhrmann [1997] for a recent study). However, I will not elaborate upon this connection here.

phism from $\mathcal{M}$ into $\mathcal{M}'$. A similar notion of embeddability can be defined also when the first term of the homomorphism is the language $\mathcal{L}$ itself rather than one of its models. This can be done simply by replacing '**s**', '**g**', and '**E**' for '**d**', '**h**', and '**I**' in definition (c) above. The resulting generalization will be very convenient in the sequel, and I shall use the notation '$\sqsubseteq$' and '$\sqsupseteq$' indifferently to designate a relation of embeddability between two models or between a language and a model. The term 'homomorphism', too, will be used indifferently in the two cases.

1.3. VALUATIONS

With the above definitions of $\mathbb{L}$ and $\mathbb{M}$, the general scope of our semantic framework is finally set. Our task now is to try and build a bridge between these two classes *via* an appropriate notion of a valuation (of a language $\mathcal{L}$ on a model $\mathcal{M}$).

1.3.1. PRELIMINARIES

What is needed, intuitively, is some means of assigning values to the expressions of a given language on the basis of the information provided by its models.

Clearly, despite the abstractness and generality of our definition of $\mathbb{L}$, the problems involved in this task are due primarily to the wideness of $\mathbb{M}$. If every element of $\mathbb{M}$ were sharp, in fact, i.e., if we only considered models that comply with the standard requirements of consistency and completeness, the account would now be quite straightforward. For in that case we could simply rely on the following facts (where $\mathcal{L}$ and $\mathcal{M}$ are arbitrarily chosen):

(1) If $\mathcal{M} \in \mathit{COMP}_{\mathcal{L}}$, then there exists at least one homomorphism from $\mathcal{L}$ into $\mathcal{M}$.
(2) If $\mathcal{M} \in \mathit{CONS}_{\mathcal{L}}$, then there exists at most one homomorphism from $\mathcal{L}$ into $\mathcal{M}$.
(3) If $\mathcal{M} \in \mathit{SHRP}_{\mathcal{L}}$, then there exists exactly one homomorphism from $\mathcal{L}$ into $\mathcal{M}$.

[*Proof.* Assume that $\mathcal{M}\in COMP_{\mathcal{L}}$. Then $\mathcal{D}R=C_{\mathcal{M}}$ for all $R\in\mathcal{R}\mathbf{d}$, so we can find a function f_0: $\mathcal{R}\mathbf{s}\rightarrow\bigcup\mathcal{R}\mathbf{I}$ that satisfies the general condition $f_0(\mathbf{s}(\beta,t))\subseteq\mathbf{d}(\beta,t)$ for all $\langle\beta,t\rangle\in\mathcal{D}\mathbf{s}$ (recall that $\mathbf{s}$ is one-one). Moreover, $\mathcal{D}R=C_{\mathcal{M}}$ for all $R\in\mathcal{R}\mathbf{h}$. Thus, since $\mathbf{g}$ is one-one and well-grounded on $\mathcal{D}\mathbf{s}$, by the recursion theorem we can extend f_0 to a complete mapping f: $\bigcup\mathcal{R}\mathbf{E}\rightarrow\bigcup\mathcal{R}\mathbf{I}$ that satisfies the condition $f(\mathbf{g}(x,y))\subseteq\mathbf{h}(f(x),f(y))$ for all $\langle x,y\rangle\in\mathcal{D}\mathbf{g}$. By definition such a function is a homomorphism from $\mathcal{L}$ into $\mathcal{M}$, which proves (1). As for (2), assume that $\mathcal{M}\in CONS_{\mathcal{L}}$. Then each element of $\mathcal{R}\mathbf{d}\cup\mathcal{R}\mathbf{h}$ is a function, and the identity of any pair of homomorphisms f and f' can be certified inductively. Pick any $x\in EXP_{\mathcal{L}}$ and suppose that $f(y)=f'(y)$ for every $y\in EXP_{\mathcal{L}}$ such that $\lambda_{\mathcal{L}}(y)<\lambda_{\mathcal{L}}(x)$. If $x=\mathbf{s}(\beta,t)$, then we have the identities $f(x)=\mathbf{d}(\beta,t)=f'(x)$ by definition of homomorphism, whereas if $x=\mathbf{g}(x,y)$, then we are sure to have the identities $f(x)=\mathbf{h}(f(y),f(z))=\mathbf{h}(f'(y),f'(z))=f'(x)$ by inductive hypothesis (since $\lambda_{\mathcal{L}}(y),\lambda_{\mathcal{L}}(z)<\lambda_{\mathcal{L}}(\mathbf{g}(y,z))$). This shows that $f=f'$. Given (1) and (2), fact (3) is now an immediate consequence of the definition of $SHRP_{\mathcal{L}}$.]

What this means, effectively, is that given a language $\mathcal{L}$, any model $\mathcal{M}\in SHRP_{\mathcal{L}}$ supplies all the information that is needed in order to associate every expression in $\bigcup\mathcal{R}\mathbf{E}$ with a suitable interpretation in $\bigcup\mathcal{R}\mathbf{I}$, hence with a unique semantic value in each relevant context $c\in C_{\mathcal{M}}$: the function $\mathbf{d}$ assigns an interpretation to the basic expressions, and $\mathbf{h}$ tells us how to compute the interpretation of a compound expression given the interpretations of its component parts. Since all this information is perfectly reflected in the unique homomorphism that relates $\mathcal{L}$ to $\mathcal{M}$, such a function (call it $\mathcal{V}$) would therefore be the natural candidate for the role of a valuation of $\mathcal{L}$ on $\mathcal{M}$, yielding the obvious values:

if $x=\mathbf{s}(\beta,t)$, then $\mathcal{V}(x)=\mathbf{d}(\beta,t)$
if $x=\mathbf{g}(y,z)$, then $\mathcal{V}(x)=\mathbf{h}(\mathcal{V}(y),\mathcal{V}(z))$.

I take this to be the right account—the standard account—in the case of sharp models: the value of a compound expression in any context is determined entirely by the values of its component parts.[27] It is clear,

---

[27] The underlying principle is the one originally stated in Frege [1892] and developed in Tarski [1933] for certain types of elementary languages.

however, that this account cannot be immediately extended to cover every other case. For, in general, the converses of (1)–(3) hold too:

(1') There exists at least one homomorphism from $\mathcal{L}$ into $\mathcal{M}$ only if $\mathcal{M} \in \mathsf{COMP}_{\mathcal{L}}$.
(2') There exists at most one homomorphism from $\mathcal{L}$ into $\mathcal{M}$ only if $\mathcal{M} \in \mathsf{CONS}_{\mathcal{L}}$.
(3') There exists exactly one homomorphism from $\mathcal{L}$ into $\mathcal{M}$ only if $\mathcal{M} \in \mathsf{SHRP}_{\mathcal{L}}$.

[*Proof.* (1') is an immediate consequence of the definition of homomorphism in 1.2.5. For assume that $\mathcal{M} \notin \mathsf{COMP}_{\mathcal{L}}$ and let $f: \bigcup\mathcal{R}\mathbf{E} \to \bigcup\mathcal{R}\mathbf{I}$. If $\mathbf{d}(\beta,t)[c]=\varnothing$ for some $\langle\beta,t\rangle \in \mathcal{D}\mathbf{s}$ and some $c \in C_{\mathcal{M}}$, then the condition $f(\mathbf{s}(\beta,t)) \subseteq \mathbf{d}(\beta,t)$ will obviously fail, whereas if $\mathbf{h}(f(x),f(y))[c]=\varnothing$ for some $\langle x,y\rangle \in \mathcal{D}\mathbf{g}$, then it is the condition $f(\mathbf{g}(x,y)) \subseteq \mathbf{h}(f(x),f(y))$ that will fail. Therefore, if $\mathcal{M} \notin \mathsf{COMP}_{\mathcal{L}}$, f cannot generally qualify as a homomorphism from $\mathcal{L}$ into $\mathcal{M}$. (There is still the possibility that the only source of incompleteness lies in some element of $\mathcal{R}\mathbf{h}-(\mathcal{R}f \times \mathcal{R}f)$. In that case, f may turn out to be homomorphic even if $\mathcal{M} \notin \mathsf{COMP}_{\mathcal{L}}$. However, for the purpose of the present argument such exceptions are uninteresting and can safely be disregarded.) As for (2'), assume that $\mathcal{M} \notin \mathsf{CONS}_{\mathcal{L}}$ and suppose that f is a homomorphism from $\mathcal{L}$ into $\mathcal{M}$. If $\mathbf{d}(\beta,t)[c] \not\approx 1$ for some $\langle\beta,t\rangle \in \mathcal{D}\mathbf{s}$ and some $c \in C_{\mathcal{M}}$, then there must exist a mapping g_0: $\mathcal{R}\mathbf{s} \to \mathcal{R}\mathbf{d}$ such that $\mathrm{g}_0(\mathbf{s}(\beta,t))(c) \in \mathbf{d}(\beta,t)[c]-\{f(\mathbf{s}(\beta,t))(c)\}$, and such a mapping will generate a homomorphism g distinct from f. Likewise, if g_0 is unique but $\mathbf{h}(f(x),f(y))[c] \not\approx 1$ for some $\langle x,y\rangle \in \mathcal{D}\mathbf{g}$, consider a mapping h_0 that extracts a total function $\mathrm{h}_0(a,b) \subseteq \mathbf{h}(a,b)$ for each $a,b \in \mathcal{D}\mathbf{h}$. (The existence of such a function is ensured by the axiom of choice.) Then, again, h_0 will generate a homomorphism distinct from f whenever $\mathrm{h}_0(f(y),f(z))(c) \neq f(f(y),f(z))(c)$, which means that f will generally not be unique. (Again, there may be exceptions when the only source of inconsistency of $\mathcal{M}$ lies in some element of $\mathcal{R}\mathbf{h}-(\mathcal{R}f \times \mathcal{R}f)$, but these are negligible cases.) This establishes (2'). Fact (3') now follows immediately from (1')–(2') by definition of $\mathsf{SHRP}_{\mathcal{L}}$, with the obvious negligible exceptions.]

Intuitively, the point here is that if $\mathcal{M} \notin \mathsf{COMP}_{\mathcal{L}}$, then the information supplied by $\mathcal{M}$ is *not enough* to evaluate every $\mathcal{L}$-expression in every

$\mathcal{M}$-context (at least, there is no obvious way of doing that on the basis of the approximate values delivered by **d** and **h**), whereas if $\mathcal{M} \notin CONS_{\mathcal{L}}$, then we have *too much* information to determine a unique evaluation (of some $\mathcal{L}$-expressions in some $\mathcal{M}$-contexts). This situation reflects one important sense in which $MOD_{\mathcal{L}}$ is wider than $SHRP_{\mathcal{L}}$: the requirements of totality and functionality—as classically imposed on the outputs of **d** and **h**—are not included in our definition of a model $\mathcal{M}$. Hence we cannot expect a model to induce a complete and unique valuation directly on the basis of those outputs. For instance, if the denotation of a symbol x is merely an approximate member of the domain of interpretation, then there is no way **h** can use it to determine the value of an expression containing x.[28] This is what I meant at the beginning of the present section: the difficulties involved in our task are not due to the abstractness of our definition of $\mathbb{L}$, but to the wideness of our characterization of $\mathbb{M}$ (i.e., of $MOD_{\mathcal{L}}$ for $\mathcal{L} \in \mathbb{L}$). Indeed, we see that the problem has a straightforward solution whenever $\mathcal{M} \in SHRP_{\mathcal{L}}$, regardless of the structure of $\mathcal{L}$.

On this basis, the second important thing to notice concerns the structure of $\mathbb{M}$ itself. In section 1.2.5 it was remarked that $\mathbb{M}$ is partially ordered by the extension relation $\sqsubseteq$ (or $\sqsupseteq$). This is a nice property insofar as it justifies a natural reading of $\sqsubseteq$ as an ordering going up-hill in terms of degrees of definiteness. But there is more to it. For given any language $\mathcal{L}$ and any context set C, $\sqsubseteq$ turns out to be a lattice ordering of $MOD_{\mathcal{L},C}$. This means that if we take any set of commensurate models for a given language $\mathcal{L}$ and put them together, either by $\bigsqcup$ (the lattice join) or by $\bigsqcap$ (the lattice meet),[29] we still get a model for $\mathcal{L}$. More specifically, the following identities hold:

(4) $COMP_{\mathcal{L},C}$ is the closure of $SHRP_{\mathcal{L},C}$ under $\bigsqcup$;
(5) $CONS_{\mathcal{L},C}$ is the closure of $SHRP_{\mathcal{L},C}$ under $\bigsqcap$;
(6) $MOD_{\mathcal{L},C}$ is the closure of $SHRP_{\mathcal{L},C}$ under both $\bigsqcap$ and $\bigsqcup$.

---

[28] There would be no difficulty if **h** were so characterized as to include approximate interpretations in its domain. But that would amount to construing $\bigcup\mathcal{R}\mathbf{I}$ itself as a set of approximate interpretations, and we have seen that such a course would not yield a genuine generalization of standard semantics. Construing $\bigcup\mathcal{R}\mathbf{I}$ as a set of approximate interpretations is essentially equivalent to construing $\mathcal{M}$ as a relational model. And all relational models are sharp.

[29] For simplicity, I write $\bigsqcup$ and $\bigsqcap$ instead of $\bigsqcup_{\sqsubseteq}$ and $\bigsqcap_{\sqsubseteq}$, respectively.

[*Proof*. Consider (4). That $SHRP_{\mathcal{L},C} \subseteq COMP_{\mathcal{L},C}$ is true by definition. To see that each non-empty set $X \subseteq COMP_{\mathcal{L},C}$ has a supremum $\bigsqcup X$, define a model $\mathcal{M}^{\cup}=(\mathbf{d}^{\cup},\mathbf{h}^{\cup},\mathbf{I}^{\cup})$ by setting (i) $\mathbf{d}^{\cup}(\beta,t)=\bigcup\{\mathbf{d}(\beta,t)\colon \mathcal{M}\in X\}$ for all $\langle\beta,t\rangle\in\mathcal{D}\mathbf{d}$, (ii) $\mathbf{h}^{\cup}(x,y)=\bigcup\{\mathbf{h}(x,y)\colon \mathcal{M}\in X\}$ for all $\langle x,y\rangle\in\mathcal{D}\mathbf{h}$, and (iii) $\mathbf{I}_t^{\cup}=\bigcup\{\mathbf{I}_t\colon \mathcal{M}\in X\}$ for all $t\in\mathcal{D}\mathbf{E}$. Then it is easy to verify that $\mathcal{M}^{\cup}\in COMP_{\mathcal{L},C}$, that every element of X is extended by $\mathcal{M}^{\cup}$, and that every other model with these properties is an extension of $\mathcal{M}^{\cup}$ itself. Hence $\mathcal{M}^{\cup}=\bigsqcup X$. Finally, to see that $COMP_{\mathcal{L},C}$ is the smallest class including $SHRP_{\mathcal{L},C}$ and closed under $\bigsqcup$, suppose $K\subset COMP_{\mathcal{L},C}$ and pick an arbitrary model $\mathcal{M}\in COMP_{\mathcal{L},C}-K$. Let $X=COMP_{\mathcal{L},C}\cap\sqsupseteq[\mathcal{M}]$ and define $Y=\{\mathcal{M}'\colon \mathcal{M}'$ is $\sqsupseteq$-maximal in $X\}$. Then $(X,\sqsupseteq)$ is a partially ordered class[30] and $Y\subseteq X\subseteq COMP_{\mathcal{L},C}$. And since every chain $(Z,\sqsupseteq)$ with $Z\subseteq X$ has an $\sqsupseteq$-upper bound in X (just take the model $\mathcal{M}^{\cap}=\bigsqcap Z$, defined in analogy to $\mathcal{M}^{\cup}$ above), it also follows that $Y\neq\varnothing$ (by Zorn's Lemma). But obviously $Y\subseteq CONS_{\mathcal{L}}$ (by the requirement of $\sqsupseteq$-maximality). Thus $Y\subseteq SHRP_{\mathcal{L},C}$, and inspection shows that $\mathcal{M}=\bigsqcup Y$. Now, $\mathcal{M}\notin K$. Hence, since K was arbitrarily chosen, it follows that $COMP_{\mathcal{L},C}$ is indeed the smallest class with the desired properties. This establishes (4). The proof of (5) is perfectly dual, whence (6) follows easily.]

These properties of $\mathbb{M}$ have some intrinsic interest. More important, however, they suggest that the class of all models for any given language may be regarded as being generated from an initial class of sharp models by combining them in some way. In other words, (4)–(6) suggest that any non-sharp model $\mathcal{M}$ may be conceived of in terms of a class of commensurate models which are perfectly sharp and which—together—contain the same amount of information as $\mathcal{M}$. In fact, for any given model $\mathcal{M}$ we can even provide an exact specification of the class of its sharp "ancestors". To this end, let a model $\mathcal{M}'$ count as a *constriction* of a model $\mathcal{M}$ (in short, $\mathcal{M}'\preccurlyeq\mathcal{M}$) if and only if $\mathcal{M}'$ is a $\sqsubseteq$-maximal consistent restriction of $\mathcal{M}$, and let $\mathcal{M}'$ count as a *completion* of $\mathcal{M}$ (in short, $\mathcal{M}'\succcurlyeq\mathcal{M}$) if and only if it is a $\sqsubseteq$-minimal complete extension of $\mathcal{M}$. Intuitively, a constriction of a given model $\mathcal{M}$ may be regarded as one possible way of "weeding out" every glut of $\mathcal{M}$ (if any), while a completion of $\mathcal{M}$ may be regarded as one possible way of

---

[30] I write $(X,\sqsupseteq)$ as short for $(X,\sqsupseteq\restriction X)$.

"filling in" all of its gaps (if any).[31] The relevant facts can then be stated as follows:

(4') $\mathcal{M}\in \mathit{COMP}_{\mathcal{L},C}$ if and only if $\mathcal{M}=\bigsqcup\{\mathcal{M}': \mathcal{M}'\preccurlyeq\mathcal{M}\}$;
(5') $\mathcal{M}\in \mathit{CONS}_{\mathcal{L},C}$ if and only if $\mathcal{M}=\bigsqcap\{\mathcal{M}': \mathcal{M}'\succcurlyeq\mathcal{M}\}$;
(6') $\mathcal{M}\in \mathit{MOD}_{\mathcal{L},C}$ if and only if $\mathcal{M}=\bigsqcup\{\bigsqcap\{\mathcal{M}': \mathcal{M}'\succcurlyeq\mathcal{M}''\}: \mathcal{M}''\preccurlyeq\mathcal{M}\}$

where each $\mathcal{M}'$ is sure to be an element of $\mathit{SHRP}_{\mathcal{L},C}$.

[*Proof.* Consider (4'). From right to left the implication is trivial, as completeness is preserved under extensions. Conversely, suppose that $\mathcal{M}\in \mathit{COMP}_{\mathcal{L},C}$ and let $X=\{\mathcal{M}': \mathcal{M}'\preccurlyeq\mathcal{M}\}$. By Zorn's Lemma $X\neq\varnothing$, and by definition every $\mathcal{M}'\in X$ is a consistent restriction of $\mathcal{M}$ with $C_{\mathcal{M}'}=C_{\mathcal{M}}=C$. So $X\subseteq \mathit{SHRP}_{\mathcal{L},C}$. Hence we only need to verify that any model extending each $\mathcal{M}'\in X$ is an extension of $\mathcal{M}$ itself. Let $\mathcal{M}''$ be such a model and consider $\langle\beta,t\rangle\in\mathcal{D}\mathbf{s}$, $\langle x,y\rangle\in\mathcal{D}\mathbf{h}$, and $t\in\mathcal{D}\mathbf{I}$. Clearly we have $\mathbf{d}(\beta,t)\subseteq\mathbf{d}''(\beta,t)$, otherwise there would exist $c\in C$ and $u\in\bigcup\mathcal{R}\mathbf{D}$ such that $\langle c,u\rangle\in\mathbf{d}(\beta,t)-\mathbf{d}''(\beta,t)$, and the model obtained from any element $\mathcal{M}'\in X$ by replacing $\mathbf{d}'(\beta,t)$ with $\mathbf{d}'(\beta,t)(c/u)$ would be extended by $\mathcal{M}$ but not by $\mathcal{M}''$, contrary to our assumption. Similarly, we must have $\mathbf{h}(x,y)\subseteq\mathbf{h}''(x,y)$, for otherwise we could find $c\in C$ and $u\in\bigcup\mathcal{R}\mathbf{D}$ such that $\langle c,u\rangle\in\mathbf{h}(x,y)-\mathbf{h}''(x,y)$, and the model obtained from any element $\mathcal{M}'\in X$ by replacing $\mathbf{h}'(x,y)$ with $\mathbf{h}'(x,y)(c/u)$ would contradict our assumption. Finally, we must also have $\mathbf{I}_t\subseteq\mathbf{I}_t''$, otherwise we would have $\mathbf{I}_t'\not\subseteq\mathbf{I}_t''$ for some $\mathcal{M}'\in X$, violating our assumption. Thus $\mathcal{M}\sqsubseteq\mathcal{M}''$ and the desired result follows by generalization. The proof of (5') is perfectly parallel, whence (6') follows immediately given that completions and constrictions are always complete and consistent, respectively.]

---

[31] In view of note 26, one could observe that this intuitive reading is in some respects inadequate. For instance, if $\mathcal{M}$ is an incomplete stratified model interpreting a functor x in a context c as a partial relation R, a completion $\mathcal{M}'$ of $\mathcal{M}$ is not a stratified model where the denotation of x in c is a total relation $R'\supseteq R$, but a non-stratified model where the denotation of x in c is a partial relation (R itself) that *behaves* totally (relative to $\mathbf{h}'$). Similarly, according to our definition a constriction of an inconsistent stratified model $\mathcal{M}$ is a non-stratified model $\mathcal{M}'$ where relations *behave* functionally (relative to $\mathbf{h}'$). Once again, I take this to be a minor issue: one can always define a correspondence between models that are completions (constrictions) in our sense and models that are completions (constrictions) in the alternative sense, claiming that the former "represent" the latter. For my purposes this will suffice.

Note that these facts are a natural byproduct of the way we extended the standard notion of a model. If, say, $\mathcal{M}$ is a complete model for $\mathcal{L}$, then a symbol $\mathbf{s}(\beta,t)$ of $\mathcal{L}$ can be classified as "determined" or "over-determined" in a context $c\in C$ depending on whether $\mathbf{d}(\beta,t)[c]$ contains one or more denotations, respectively, and an ordered pair $\langle x,y\rangle\in\mathcal{D}\mathbf{h}$ can be classified as "determined" or "overdetermined" in c depending on whether $\mathbf{h}(x,y)[c]$ has one or more elements. However, to say that $\mathbf{d}(\beta,t)[c]$ contains multiple denotations is to say that $\mathbf{s}(\beta,t)$ may be regarded as having a certain denotation in c under some way of making $\mathbf{d}(\beta,t)$ more precise, a different denotation in c under some other way of making $\mathbf{d}(\beta,t)$ precise, and so on. Each way of making $\mathbf{d}(\beta,t)$ more precise with respect to c corresponds to a way of weeding out a glut in $\mathbf{d}(\beta,t)$; and since the same account applies to c as well as to any other context of $\mathcal{M}$, this means that $\mathbf{d}(\beta,t)$ itself may be regarded as tantamount to a class of glutless denotation assignments—a class of functions $f(\beta,t)\colon C\rightarrow\mathbf{D}_t$ whose union equals $\mathbf{d}(\beta,t)$. Likewise, to say that a pair $\langle x,y\rangle\in\mathcal{D}\mathbf{h}$ is overdetermined in a context $c\in C$ is to say that it may be regarded as yielding a certain value in c under some way of making $\mathbf{h}(x,y)$ more precise, a different value under some other way of making $\mathbf{h}(x,y)$ precise, and so on. Hence $\mathbf{h}(x,y)$ itself may be regarded as tantamount to a class of functions—a class whose union is $\mathbf{h}(x,y)$. By generalization, thinking of $\mathbf{d}$ and $\mathbf{h}$ in this way amounts to interpreting $\mathcal{L}$ by means of a class of complete *and* consistent models—the class of $\mathcal{M}$'s constrictions. And to say that $\mathcal{M}$ is the join of such models is just the correct way to express this fact—i.e., (4') amounts to the identities

$$\text{for all } \langle\beta,t\rangle\in\mathcal{D}\mathbf{s}\text{:}\quad \mathbf{d}(\beta,t) = \bigcup\{\mathbf{d}'(\beta,t)\colon \mathcal{M}'\preccurlyeq\mathcal{M}\}$$
$$\text{for all } \langle x,y\rangle\in\mathcal{D}\mathbf{h}\text{:}\quad \mathbf{h}(x,y) = \bigcup\{\mathbf{h}'(x,y)\colon \mathcal{M}'\preccurlyeq\mathcal{M}\}.$$

Similarly, interpreting $\mathcal{L}$ by means of a consistent model $\mathcal{M}$ amounts to interpreting $\mathcal{L}$ by means of a class of consistent *and* complete models, each of which corresponds to some way of making the components of $\mathcal{M}$ more precise. This is the class of $\mathcal{M}$'s completions; and to say that $\mathcal{M}$ is the meet of such models is the right way of saying that it is somehow "made of" them: every gap in the values of $\mathbf{d}$ and $\mathbf{h}$ (if any) is a gap in the pattern of agreement of the denotation assignments and the structural operations of those models. In other words, (5') amounts to the identities

$$\text{for all } \langle\beta,t\rangle\in\mathcal{D}\mathbf{s}:\quad \mathbf{d}(\beta,t) = \bigcap\{\mathbf{d}'(\beta,t): \mathcal{M}'\succcurlyeq\mathcal{M}\}$$
$$\text{for all } \langle x,y\rangle\in\mathcal{D}\mathbf{h}:\quad \mathbf{h}(x,y) = \bigcap\{\mathbf{h}'(x,y): \mathcal{M}'\succcurlyeq\mathcal{M}\}.$$

In general, then, interpreting $\mathcal{L}$ by means of an arbitrary model $\mathcal{M}$, possibly inconsistent and incomplete, amounts to interpreting $\mathcal{L}$ by means of a class of classes of sharp models: to each constriction of $\mathcal{M}$ there corresponds a class of completions thereof. And (6') says that the constituents of $\mathcal{M}$ can be characterized in terms of the constituents of such "sharpenings":

$$\text{for all } \langle\beta,t\rangle\in\mathcal{D}\mathbf{s}:\quad \mathbf{d}(\beta,t) = \bigcup\{\bigcap\{\mathbf{d}'(\beta,t): \mathcal{M}'\succcurlyeq\mathcal{M}''\}: \mathcal{M}''\preccurlyeq\mathcal{M}\}$$
$$\text{for all } \langle x,y\rangle\in\mathcal{D}\mathbf{h}:\quad \mathbf{h}(x,y) = \bigcup\{\bigcap\{\mathbf{h}'(x,y): \mathcal{M}'\succcurlyeq\mathcal{M}''\}: \mathcal{M}''\preccurlyeq\mathcal{M}\}.$$

There are many interesting morals one could draw from these facts. But what is most relevant here is this: apart from whatever intrinsic interest they may have, (4')–(6') provide a solid counterpart to (1')–(3'). Above we saw that valuations on sharp models are straightforward, although we cannot immediately apply the same pattern to other sorts of models. Here we see that every other sort of model can in a way be represented by a class of sharp models. Hence, we may suggest, valuations on a model that is not sharp may be construed from the relevant class of sharp models, for the latter provides the same relevant informational input.

It is precisely this suggestion—based on both sides of the coin, as it were—that I wish to exploit here. We need a notion of a valuation that is general enough to fit every model $\mathcal{M}$ of any language $\mathcal{L}$, and we have seen that this is likely to produce, for some $\mathcal{L}$-expressions x, not a total function $\mathcal{V}(x)\colon C_{\mathcal{M}}\to\mathbf{D}_{\tau_{\mathcal{L}}(x)}$ but a generic relation $\mathcal{V}(x)\subseteq C_{\mathcal{M}}\times\mathbf{D}_{\tau_{\mathcal{L}}(x)}$ yielding zero, one, or more values in $\mathbf{D}_{\tau_{\mathcal{L}}(x)}$ for each context in $C_{\mathcal{M}}$. In other words, the outputs of a valuation $\mathcal{V}$ will generally approximate membership in $\bigcup\mathcal{R}\mathbf{I}$, just as the outputs of **d** and **h**. Should we insist on the principle that the values of a compound expression are always exhaustively determined by the values of its component parts (the "compositionality principle"), a battle of intuitions could arise as to the correct formulation of the inductive clause. For instance, one possibility would be to proceed as follows:

if x=$\mathbf{s}(\beta,t)$, set $\mathcal{V}(x)$=$\mathbf{d}(\beta,t)$
if x=$\mathbf{g}(y,z)$, set $\mathcal{V}(x)$=$\bigcup\{\mathbf{h}(a,b)\colon a\eqsim\mathcal{V}(y) \text{ and } b\eqsim\mathcal{V}(z)\}$,

where '$\cong$' denotes the relation of approximate identity that holds between a function f and a relation R when $f(c) \in R[c]$ for every $c \in \mathcal{D}R$, i.e., when $f \upharpoonright \mathcal{D}R \subseteq R$. The smallest relation $\mathcal{V}$ satisfying these conditions—call it a *paramorphism* between $\mathcal{L}$ and $\mathcal{M}$—would certainly qualify as a generalization of the notion of a homomorphism, and it would be uniquely determined for all $\mathcal{M} \in MOD_{\mathcal{L}}$.[32] But apart from the fact that the inductive clause is only one among many options that may be suggested for the evaluation of compound expressions, it is not clear to what extent any of these would properly reflect the amount of *relevant* information actually provided by the model. To illustrate, suppose that $\mathcal{L}$ is a sentential language with a binary connective $\downarrow$, as in example 1.1.4(A), and suppose that $\mathcal{M}$ is a complete but inconsistent extensionally adequate 2-valued sentential model for $\mathcal{L}$ interpreting $\downarrow$ as the joint-denial connective, as in example 1.2.4(A). If $z = \mathbf{s}(\zeta, 0)$ is a sentence symbol that is overdetermined in a context $c \in C_{\mathcal{M}}$, i.e., if $\mathbf{d}(\zeta, 0)[c] = \{0,1\}$, then it is easy to verify that every admissible homomorphism $\mathcal{V}$ would agree on the value $\mathcal{V}(z \downarrow (z \downarrow z))(c) = 0$. (We know that there are at least two such homomorphisms by (1) and (2').) Yet the corresponding paramorphism would yield the multiple evaluation $\mathcal{V}(z \downarrow (z \downarrow z))[c] = \{0,1\}$. Alternative formulations would yield similar results, if the only guideline for evaluating compound expressions is the compositionality principle.

By contrast, facts (4')–(6') supply a more immediate way of resolving the issue.[33] Consider $COMP_{\mathcal{L}}$ first. We have seen that any inconsistent element of this class admits of several homomorphic valuations. None of them would qualify as *the* valuation of $\mathcal{L}$ on $\mathcal{M}$ because none of them reflects *all* the information available in $\mathcal{M}$. However, they do reflect such information if we take them collectively. We can assign a value v to an expression x in a context c if and only if v is assigned to x

---

[32] This would be in the spirit of Muskens [1989, 1995], which is in fact a generalization of the 4-valued model theory put forward by Dunn [1976] and further elaborated by Belnap [1977] and Woodruff [1984b] among others. (This in turn, represents a generalization of the 3-valued semantics of Kleene [1938].)

[33] As mentioned in the Introduction, the basic idea developed here may be seen as implementing a form of "supervaluationism", in the sense originally introduced by van Fraassen [1966a, 1966b] in connection with the semantics of free logics. See note 34 for further references.

in c by *some* homomorphic valuation. And it is exactly this insight that emerges from the above facts. As it turns out, such homomorphisms are neither more nor less than the class of homomorphisms between $\mathcal{L}$ and the constrictions of $\mathcal{M}$, which themselves are impeccably sharp, and we know from (4') that $\mathcal{M}$ is itself the join of those constrictions. Hence, the suggestion is to build upon these facts. Just as a model's components satisfy the equations

$$\mathbf{d}(\beta,t) = \bigcup\{\mathbf{d}'(\beta,t)\colon \mathcal{M}'\preccurlyeq\mathcal{M}\}$$
$$\mathbf{h}(x,y) = \bigcup\{\mathbf{h}'(x,y)\colon \mathcal{M}'\preccurlyeq\mathcal{M}\},$$

the valuation may be defined by the parallel condition

$$\mathcal{V}(x) = \bigcup\{\mathcal{V}'(x)\colon \mathcal{M}'\preccurlyeq\mathcal{M}\},$$

where each $\mathcal{V}'$ is the homomorphism between $\mathcal{L}$ and the corresponding $\mathcal{M}'$ whose existence and uniqueness are guaranteed by (3). For this function registers the informational content of the class of models which $\mathcal{M}$ is "made of", hence of $\mathcal{M}$ itself. (Note that in case $\mathcal{M}$ is already a perfectly sharp, glutless model, this procedure trivially reduces to the standard one, as $\mathcal{M}\in SHRP_{\mathcal{L}}$ implies $\{\mathcal{M}'\colon \mathcal{M}'\preccurlyeq \mathcal{M}\}=\{\mathcal{M}\}$. That is, when $\mathcal{M}$ is complete as well as consistent, $\mathcal{V}$ is just the plain homomorphism between $\mathcal{L}$ and $\mathcal{M}$ itself.)

More intuitively, the idea is that a complete, possibly inconsistent model $\mathcal{M}$ is one in which exactly that holds which holds in some one of its constrictions, each of which corresponds to some way of weeding out the initial gluts in $\mathcal{M}$. If a given expression x involves constituents that are overdetermined in some context c of $\mathcal{M}$, it is *prima facie* impossible to establish the value of x in c on the basis of $\mathcal{M}$. But we can look at the value x takes on the various constrictions of $\mathcal{M}$. If every such tentative valuation yields the same outcome—i.e., if $\mathcal{V}'(x)(c)$ is the same for all $\mathcal{M}'\preccurlyeq\mathcal{M}$—then we may conclude that the gluts of $\mathcal{M}$ are not relevant for the purpose of evaluating x in c, and we may therefore assign x the established value. Otherwise we shall give x *all* the values that we get, for there is no apparent reason to reject one value or another. For example, if $\mathcal{M}$ is the sentential model mentioned above, where the sentence symbol z=$\mathbf{s}(\zeta,0)$ has the denotations $\mathbf{d}(\zeta,0)[c]=\{0,1\}$, there will be some constrictions which determine the value $\mathcal{V}'(z)(c)=0$ while others will determine the value $\mathcal{V}'(z)(c)=1$. Hence the valuation induced by $\mathcal{M}$ will eventually assign both values to z in c, i.e., we obtain $\mathcal{V}(z)[c]=\{0,1\}$. For the same reason, a compound such as $z{\downarrow}z$ will be assigned both values, i.e., $\mathcal{V}(z{\downarrow}z)[c]=\{0,1\}$. So far this coincides with the paramorphic valuation, except that the values in $\mathcal{V}(z{\downarrow}z)[c]$ are computed, not directly from the values in $\mathcal{V}(z)[c]$, but indirectly from the series of values $\mathcal{V}'(z{\downarrow}z)(c)$ determined by the constrictions of $\mathcal{M}$. And when it comes to a sentence such as $z{\downarrow}(z{\downarrow}z)$ this difference becomes relevant: unlike the paramorphism, $\mathcal{V}$ will evaluate such a sentence as unambiguously false in c, i.e., $\mathcal{V}(z{\downarrow}(z{\downarrow}z))(c)=0$, since all constrictions will agree on the constant value $\mathcal{V}'(z{\downarrow}(z{\downarrow}z))(c)=0$. (For a concrete illustration, assume the sentence symbols of $\mathcal{L}$ are sentences of English and suppose c represents the *ersatz* world described in the Holmes stories. Then the sentence z = 'Watson limps' will be both true and false in c, due to the aforementioned discrepancy: in one of the stories Watson's war wound is located in his leg, but in another story it is located in his shoulder. For the same reason, the compound sentence $z{\downarrow}z$ (i.e., effectively, 'Watson does not limp') will be both true and false in c. But the compound $z{\downarrow}(z{\downarrow}z)$ (i.e., 'Watson neither does nor doesn't limp') will be false and only false in c, for it is false in each story, and never true.)

Let us now focus on the class of models that are consistent, $\mathit{CONS}_{\mathcal{L}}$. In this case the foregoing account does not immediately apply, for there is no point in looking at the constrictions of a consistent model. Indeed, if a model $\mathcal{M}$ is consistent but incomplete, then its constrictions will not be sharp, as incompleteness is preserved under $\sqsubseteq$. However, we may still rely on the above facts. In particular, we have seen that any model $\mathcal{M}\in \mathit{CONS}_{\mathcal{L}}$ turns out to be the *meet* of its sharp completions, characterized by the equations:

$$\mathbf{d}(\beta,t) = \bigcap\{\mathbf{d}'(\beta,t)\colon \mathcal{M}'\succcurlyeq\mathcal{M}\}$$
$$\mathbf{h}(x,y) = \bigcap\{\mathbf{h}'(x,y)\colon \mathcal{M}'\succcurlyeq\mathcal{M}\}.$$

Hence, in this case the natural way of defining valuations is by reference to the meet of the relevant homomorphisms:

$$\mathcal{V}(x) = \bigcap\{\mathcal{V}'(x)\colon \mathcal{M}'\succcurlyeq\mathcal{M}\},$$

This function registers the pattern of agreement among all the complete models $\mathcal{M}$ is "made of". Hence it reflects the informational content of

$\mathcal{M}$.[34] Exactly that holds in an incomplete model which holds in every way of filling in its gaps. (Again, note that if $\mathcal{M}$ is already gapless, hence sharp, this procedure reduces trivially to a homomorphic valuation, since $\mathcal{M} \in SHRP_{\mathcal{L}}$ implies $\{\mathcal{M}': \mathcal{M}' \succcurlyeq \mathcal{M}\} = \{\mathcal{M}\}$.)

Intuitively, this is perfectly dual to the idea outlined above. If an expression x involves some constituents with respect to which our model $\mathcal{M}$ is undetermined (in some context c) there is no obvious way to evaluate it (in c) on the basis of $\mathcal{M}$. But we can look at the value x takes on the various completions of $\mathcal{M}$, all of which are completely defined. If every such tentative valuation yields the same outcome—i.e., if the value $\mathcal{V}'(x)(c)$ is the same for all $\mathcal{M}' \succcurlyeq \mathcal{M}$—then we may conclude that the gaps of $\mathcal{M}$ are not relevant for the purpose of evaluating x in c, and

[34] It is this feature that assimilates the present account to van Fraassen's supervaluations (though in the case of inconsistent models it would be more appropriate to speak instead of "subvaluations"). Supervaluational semantics have been considered by various authors in relation to a variety of specific applications: from the treatment of *non-denoting terms* (Skyrms [1968], van Fraassen [1969, 1970b], Bencivenga [1977, 1978, 1980a, 1980b, 1981], Lambert & Bencivenga [1986], Bencivenga, Lambert & van Fraassen [1986]) to the treatment of *vague predicates* (Thomason [1973], Fine [1975], Kamp [1975, 1981a], Klein [1980], Pinkal [1983, 1984], Burgess & Humberstone [1987], Chierchia & McConnell-Ginet [1991], McGee & McLaughlin [1994]), *sortal incorrectness* (Martin [1970], Thomason [1972], Bergmann [1977]), *truth ascriptions* (van Fraassen [1970a], Burgess [1986], Fitting [1986], Cantini [1990, 1996], McGee [1991]), *tensed statements* (Thomason [1970], van Bendegem [1993], Restall [1995]), *propositional attitudes* (Vergauwen [1984]), as well as various issues in the philosophy of science (Lambert [1969]), metaphysics (Jackson & Pargetter [1988]), and game theoretic semantics (Blinov [1994]). (Similar ideas may be found, more or less explicitly, also in Mehlberg [1956], § 29, Przełęcki [1964, 1969, 1976, 1980], Lewis [1970, 1978, 1979], Grant [1974], and Dummett [1975].) All of these works share the same insight, viz., that of a valuation registering the pattern of agreement among a family of classical valuations, though the actual implementation may vary. (The approach followed here would be in the spirit of what Herzberger [1982] calls model-based extrapolation methods for construing supervaluations.) In any case, these works have all focused on models that are incomplete but consistent, and typically only models for elementary languages. By contrast, a parallel treatment of inconsistent models has not been given much thought in the literature (see Varzi [1991, 1994c]). Some hints can be found in Belnap [1977], Rescher [1979], Rescher & Brandom [1980], Lewis [1982, 1983b], Visser [1984], and Hyde [1997] The non-adjunctive logics pioneered by Jaśkowski [1948] can be seen in this perspective too; compare Da Costa & Dubikajtis [1977], Kotas & Da Costa [1979], Jennings & Schotch [1984], and Schotch & Jennings [1989].

we may give x the value in question. Otherwise we leave x undefined in c, for there seems to be no way to *choose* one value over the other(s)—the model does not determine it. (To illustrate, if c is a context that reflects, not the Holmes stories collectively, but one of them individually, then there are English sentences that lack a truth value in c, e.g., 'Watson was born on a Tuesday'. A compound involving such a sentence, however, may receive a definite truth value if the gap turns out to be irrelevant. The sentence 'Watson was born on a Tuesday and died the next Wednesday' comes out false no matter how the story is completed, hence we can deny it in spite of our lack of information—or so I am suggesting.)

At this point, it is not difficult to see how the two complementary suggestions outlined above can be combined and generalized so as to cover every case, including models that are neither consistent nor complete. If $\mathcal{M}$ is such a model, its constrictions are not sharp, and neither are its completions. However, the completions of its constrictions *are* complete and consistent, hence sharp, and we know from (6') that every model $\mathcal{M}$ can be represented by a class of such sharp models through the basic identities

$$\mathbf{d}(\beta,t) = \bigcup\{\bigcap\{\mathbf{d}'(\beta,t)\colon \mathcal{M}'\succcurlyeq\mathcal{M}''\}\colon \mathcal{M}''\preccurlyeq\mathcal{M}\}$$
$$\mathbf{h}(x,y) = \bigcup\{\bigcap\{\mathbf{h}'(x,y)\colon \mathcal{M}'\succcurlyeq\mathcal{M}''\}\colon \mathcal{M}''\preccurlyeq\mathcal{M}\}.$$

Hence, we may rely on this fact to build the desired bridge between languages and models in a most general form, setting

$$\mathcal{V}(x) = \bigcup\{\bigcap\{\mathcal{V}'(x)\colon \mathcal{M}'\succcurlyeq\mathcal{M}''\}\colon \mathcal{M}''\preccurlyeq\mathcal{M}\}.$$

In other words: if there is some constriction of $\mathcal{M}$ every completion of which assigns a value v to an expression x in a context c (by the relevant homomorphism), then set $v\in\mathcal{V}(x)[c]$; otherwise set $v\notin\mathcal{V}(x)[c]$. Exactly that holds in a model $\mathcal{M}$ which holds in any one of its constrictions; and exactly that holds in a constriction which holds in each one of its completions.

With this, the basic idea is fixed. The resulting notion of a valuation is not standard and does not satisfy the compositionality principle. However, it is standard (and it satisfies that principle) *at a remove*: the values of a given expression on a given model $\mathcal{M}$ are determined by the values of that expression on the sharp models that sort out the gaps and gluts of $\mathcal{M}$; and these latter values are unproblematic insofar as they are computed in a standard way. The semantics of an expression, we may also say, is a function not only of what the model says *explicitly*, but also of what it says *implicitly*—not only of the actual interpretation of the expression, which may be inadequate, but also of its potential interpretations, corresponding to the possibilities of sharpening its actual interpretation.[35] Before certifying the above suggestion as a final definition, however, there are two important factors that still need to be clarified. Both are somehow involved in the intuitive understanding of the *process* whereby a given model is constricted and/or completed (for the purpose of evaluating an expression).

First of all, as I have characterized it, this process is entirely based on the lattice-theoretic properties of $\mathbb{M}$, i.e., on (4')–(6'). Yet one could observe that the relevant notions of completion and constriction are exceedingly strict. To evaluate an $\mathcal{L}$-expression x one need not consider models that provide complete and consistent information relative to each and every $\mathcal{L}$-expression. Ideally, one only need consider models that are complete and consistent *relative* to x. For instance, our schema requires that in order to evaluate a sentence involving the overdefined component 'Watson limps' in the world of Sherlock Holmes, we consider models that are complete and consistent relative to the entire language. But there is no obvious reason why our task should require a radical operation like that—an operation that in some cases is practically impossible. A more natural account would be to consider sharpenings that are complete and consistent relative to the sentence to be evaluated—models in which those facts are clarified that are relevant to the evaluation of the sentence 'Watson limps'. Other facts need not be investigated. Similar considerations apply in case of a gap: Why should the task of evaluating a simple component such as 'Watson was born on a Tuesday' require a global grasping of the indefinitely many sharpenings of the entire world of Sherlock Holmes? It should suffice to consider models that are complete and consistent relative to Watson's date of birth.

---

[35] The terminology is from Fenstad [1997] and corresponds to a suggestion of Fine [1975], p. 277. See Sanford [1976], Tye [1989], Lehmann [1994], and Fodor & LePore [1996] for some extensional misgivings; Bencivenga [1997] and Morreau [1998] for replies.

To express this precisely, where $x \in EXP_{\mathcal{L}}$ let an *x-homomorphism* from $\mathcal{L}$ to $\mathcal{M}$ be any map $f: \omega_{\mathcal{L}}(x) \rightarrow \bigcup \mathcal{R}\mathbf{I}$ such that, for all constituents $w \in \omega_{\mathcal{L}}(x)$ and all $\beta, t, y,$ and z as appropriate:

$$\text{if } w=\mathbf{s}(\beta,t), \text{ then } f(w) \subseteq \mathbf{d}(\beta,t)$$
$$\text{if } w=\mathbf{g}(y,z), \text{ then } f(w) \subseteq \mathbf{h}(f(y),f(z)).$$

Intuitively, if there exists at least one such map, then $\mathcal{M}$ is *x-complete* for $\mathcal{L}$, while if there exists at most one x-homomorphism, $\mathcal{M}$ is *x-consistent*. Then the point is that, in general, to evaluate a given expression x we only need consider restrictions and extensions that are *x-sharp*, i.e., x-complete and x-consistent.[36] (The classes of models with these properties will be written as $COMP_{\mathcal{L}}(x)$, $CONS_{\mathcal{L}}(x)$, and $SHRP_{\mathcal{L}}(x)$, respectively.)

As it turns out, this change of attitude would not by itself determine any difference in the final outcome. Every sharp model is x-sharp; and for every model $\mathcal{M}'$ that qualifies as an x-complete extension of an x-consistent restriction of a given model $\mathcal{M}$ one can always find a model $\mathcal{M}''$ that qualifies as a full completion of a full constriction of $\mathcal{M}$ and which agrees with $\mathcal{M}'$ as far as x is concerned, i.e., such that the value of x under the homomorphism induced by $\mathcal{M}''$ coincides with the value of x under the x-homomorphism induced by $\mathcal{M}'$. However, things will be different once we consider a second important limitation of the notion of a valuation suggested above. As it stands, the account requires that in order to evaluate a given $\mathcal{L}$-expression x we consider the value of x in *every* sharpening of the given model (or at least in every x-sharpening, as we may now say). This is obviously too strict. If only models of a certain class $K \subseteq MOD_{\mathcal{L}}$ are regarded as "admissible", then one might want to consider only sharpenings that belong to K: only admissible models should qualify as admissible sharpenings. For instance, if the model contains a world c which is meant to reflect the Holmes stories, then Watson's date of birth is not specified in c. But an extension which says that Watson was born in 1725 would be too far-fetched to be included in the admissible sharpenings of that model. Likewise, some extensions should be left out on account of certain "penumbral connec-

---

[36] The notion of an x-sharp model may be viewed as a generalization of the notion of an "A-world" (A a sentence) introduced in Bencivenga [1977].

tions"[37] which are determined by the model in spite of its gaps. Watson's and Lestrade's dates of birth are both left undefined. But suppose the model says that Watson is older than Lestrade. Then an extension in which Watson was born in 1850 and Lestrade in 1849 should not be admissible. Analogous considerations apply to glut-deleting sharpenings. If both Mycroft and Lestrade are inconsistently modeled in c as tall and also as short, and if the model says explicitly that Mycroft is taller than Lestrade, then a consistent restriction in which Mycroft is short and Lestrade tall is simply not admissible.

More generally, the point is that one might want to assume a suitable relation R to be defined on $MOD_{\mathcal{L}}$ so that only models from the class $R[\mathcal{M}]$ would qualify as admissible sharpenings of a given $\mathcal{M} \in MOD_{\mathcal{L}}$. From this point of view, the account suggested so far would correspond to the limit case where every model is R-related to every model, i.e., where $R = MOD_{\mathcal{L}} \times MOD_{\mathcal{L}}$; but other cases should be allowed as well. This will obviously make a difference with regard to the evaluation of certain expressions, for it will affect the number of sharpenings to be compared. And, consequently, the first emendment considered above will make a difference too. A model may have an admissible x-sharpening that does not agree with any admissible *full* sharpenings. And it may have admissible x-sharpenings even if it has no admissible full sharpenings at all.

To attain the desired degree of generality, our notion of valuation needs therefore to be supplemented by some means of accounting for both these factors: some means of relativizing sharpenings to expressions, and some means of specifying the desired relation of admissibility among models. In particular, this entails that a model will not, by itself, determine a unique valuation of the language: it will do so only with respect to a given admissibility relation. Thus, where $x \in EXP_{\mathcal{L}}$ and $R \subseteq MOD_{\mathcal{L}} \times MOD_{\mathcal{L}}$, let a model $\mathcal{M}'$ count as an *(x,R)-constriction* of a model $\mathcal{M}$ (written $\mathcal{M}' \preccurlyeq^x_R \mathcal{M}$) if and only if $\mathcal{M}'$ is a restriction of $\mathcal{M}$ which is $\sqsubseteq$-maximal in $CONS_{\mathcal{L}}(x) \cap R[\mathcal{M}]$, and let $\mathcal{M}'$ count as an *(x,R)-completion* of $\mathcal{M}$ ($\mathcal{M}' \succcurlyeq^x_R \mathcal{M}$) if and only if $\mathcal{M}'$ is an extension of

[37] This notion of "penumbral connection" is due to Fine [1975]. See also Kamp [1975, 1981a]. The general point can be traced back to van Fraassen's own generalization of supervaluations in [1966b, 1969].

$\mathcal{M}$ which is $\sqsubseteq$-minimal in $COMP_{\mathcal{L}}(x) \cap R[\mathcal{M}]$. Then the final formulation I propose is the following.

1.3.2. DEFINITION

Let $\mathcal{L}$ be a language and $\mathcal{M}$ a model for $\mathcal{L}$. A function $\mathcal{V}$ is a *valuation* of $\mathcal{L}$ on $\mathcal{M}$ if and only if it assigns each expression $x \in \bigcup \mathcal{R}\mathbf{E}$ an approximate interpretation $\mathcal{V}(x) \in \bigcup \mathcal{R}\mathbf{I}$ defined thus:

$$\mathcal{V}(x) = \bigcup \{ \bigcap \{ \mathcal{V}'(x) \colon \mathcal{M}' \succcurlyeq^x_R \mathcal{M}'' \} \colon \mathcal{M}'' \preccurlyeq^x_R \mathcal{M} \},$$

where $R \subseteq MOD_{\mathcal{L}} \times MOD_{\mathcal{L}}$ and each $\mathcal{V}'$ is the unique x-homomorphism from $\mathcal{L}$ to the corresponding $\mathcal{M}'$. In other words, $\mathcal{V}(x)$ is a relation such that, for every $c \in C_{\mathcal{M}}$ and every $v \in \mathbf{D}_{\tau_{\mathcal{L}}(x)}$, $v \in \mathcal{V}(x)[c]$ if and only if $v = \mathcal{V}'(x)(c)$ for each (x,R)-completion $\mathcal{M}'$ of some one (x,R)-constriction $\mathcal{M}''$ of $\mathcal{M}$. The relation $\mathcal{V}(x)$ is called the *value assignment* for x in $\mathcal{M}$ (under R), and the elements of $\mathcal{V}(x)[c]$ are called the *values* of x in c (under R).

1.3.3. REMARKS

For definiteness, I will assume that the valuations of a given language $\mathcal{L}$ are defined only for expressions of $\mathcal{L}$. Thus, every admissibility relation $R \subseteq MOD_{\mathcal{L}} \times MOD_{\mathcal{L}}$ will induce exactly one valuation—call it the *R-valuation*—for each $\mathcal{M} \in MOD_{\mathcal{L}}$. I have not imposed any restrictions on such relations. Ideally, we shall only be interested in admissibility relations for which the identity

$$\mathcal{M} = \bigsqcup \{ \bigsqcap \{ \mathcal{M}' \colon \mathcal{M}' \succcurlyeq^x_R \mathcal{M}'' \} \colon \mathcal{M}'' \preccurlyeq^x_R \mathcal{M} \}$$

holds for every $x \in EXP_{\mathcal{L}}$ and every $\mathcal{M} \in MOD_{\mathcal{L}}$. An R satisfying this condition will be called an *ideal* admissibility relation on $MOD_{\mathcal{L}}$. But for the sake of generality I shall leave open the possibility that any other sort of relation (for instance, an arbitrary partial ordering) will determine a valuation.

As for the notation, note that there is no straight correspondence among these notions: different models may induce the same R-valuation of a given language; and different languages may have the same R-valuation on a common model. (This is possible if the languages are similar, in the sense of 1.1.5(a), and if their structural operations generate the same sets of expressions.) I will not introduce any devices to avoid ambiguities systematically, but I shall often use subscripts and superscripts in such a way as to suggest the intended interpretation. Thus, when a single language $\mathcal{L}$ is at issue, I will write '$\mathcal{V}_R$', '$\mathcal{V}'_R$', '$\mathcal{V}^R_j$', etc., for the R-valuations on the models $\mathcal{M}$, $\mathcal{M}'$, $\mathcal{M}_j$, etc. (respectively), omitting the index 'R' only if the context permits it. Also, given a model $\mathcal{M}$ and a context $c \in C_{\mathcal{M}}$, it will often be convenient to speak of the *c-segment* of a valuation $\mathcal{V}_R$ on $\mathcal{M}$. This will be denoted by '$\mathcal{V}_{R,c}$' and is defined by setting

$$\mathcal{V}_{R,c}(x) = \mathcal{V}_R(x)[c],$$

We may think of $\mathcal{V}_{R,c}$ as a local valuation registering the behavior of $\mathcal{V}_R$ in c: it associates each expression x with the set of its values in c under $\mathcal{V}_R$. And it follows from definition 1.3.2 that

$$\mathcal{V}_{R,c}(x) = \bigcup\{\bigcap\{\mathcal{V}'_{R,c}(x) \colon \mathcal{M}' \succcurlyeq^x_R \mathcal{M}''\} \colon \mathcal{M}'' \preccurlyeq^x_R \mathcal{M}\}.$$

Again, I shall use superscripts as needed and I shall omit the index 'R' when this does not impair—or improves—readability.

A few additional remarks. First of all, note that valuations are subject to a threefold partition, just as the corresponding models. A valuation $\mathcal{V}$ on a model $\mathcal{M}$ is naturally classified as complete, consistent, or sharp according to whether $\mathcal{V}(x)[c]$ contains at least one, at most one, or exactly one value (respectively) for each $x \in EXP_{\mathcal{L}}$ and each $c \in C_{\mathcal{M}}$. One immediately verifies that if a model $\mathcal{M}$ is complete, consistent, or sharp, then the same holds true of any valuation on $\mathcal{M}$—at least as long as the relevant admissibility relation R is reflexive.[38] Likewise for the relative notions of x-consistency, x-completeness, and x-sharpness. Indeed, just as models are partially ordered by the extension relation $\sqsubseteq$ defined in 1.2.5, valuations are partially ordered by the extension relation defined by

---

[38] The converse does not necessarily hold. Let $\mathcal{V}$ be a valuation on $\mathcal{M}$ and suppose that the only source of incompleteness/inconsistency in $\mathcal{M}$ lies in some element of $\mathcal{R}\mathbf{h} - (\mathcal{RV} \times \mathcal{RV})$. Then $\mathcal{V}$ will be complete/consistent even if $\mathcal{M}$ is not. Compare the remarks in the proof of (1')–(3').

$\mathcal{V} \sqsubseteq \mathcal{V}'$ if and only if $\mathcal{V}(x) \subseteq \mathcal{V}'(x)$ for all $x \in EXP_{\mathcal{L}}$.

And one verifies that whenever R is an ideal admissibility relation these orderings are R-monotonic, i.e., satisfy the conditional

if $\mathcal{M} \sqsubseteq \mathcal{M}'$ then $\mathcal{V}_R \sqsubseteq \mathcal{V}'_R$.

This is noteworthy, because it reflects our intuitive reading of these orderings as going up-hill in terms of degrees of definiteness: if $\mathcal{M}'$ involves fewer gaps than $\mathcal{M}$ (or $\mathcal{M}$ fewer gluts than $\mathcal{M}'$), then an R-valuation on $\mathcal{M}'$ will involve fewer gaps than the R-valuation on $\mathcal{M}$ (and the latter fewer gluts than the former), at least when R is ideal. Also, sharp valuations always behave homomorphically, which means that if we take R to be any reflexive relation and confine ourselves to $SHRP_{\mathcal{L}}$, then our account is faithful to the standard semantic principles discussed earlier:

if $x=\mathbf{s}(\beta,t)$, then $\mathcal{V}_R(x)=\mathbf{d}(\beta,t)$
if $x=\mathbf{g}(y,z)$, then $\mathcal{V}_R(x)=\mathbf{h}(\mathcal{V}_R(y),\mathcal{V}_R(z))$.

More generally, $\mathcal{V}_R$ will behave homomorphically with respect to any $\mathcal{L}$-expression x relative to which a given model $\mathcal{M}$ provides a complete and consistent amount of information: if $\mathcal{M} \in SHRP_{\mathcal{L}}(x)$, then $\mathcal{V}_R \upharpoonright \omega_{\mathcal{L}}(x)$ is precisely the x-homomorphism from $\mathcal{L}$ to $\mathcal{M}$. Thus, "local" sharpness does not get lost in the evaluation process.

These and other basic properties of the account will be further investigated in the next Chapter. Before illustrating the working of definition 1.3.2 in connection with some concrete examples, there is yet an important feature of the underlying approach that requires a word of comment. It concerns the general format of our definition, where the relevant model can be both incomplete and inconsistent.

Consider any language $\mathcal{L}$ along with an arbitrary admissibility relation $R \subseteq MOD_{\mathcal{L}} \times MOD_{\mathcal{L}}$. Relative to any model $\mathcal{M} \in MOD_{\mathcal{L}}$, we can define the *dual* of the corresponding R-valuation $\mathcal{V}$ in a very natural way. It is the unique relation $\mathcal{W}$ such that, for all $x \in EXP_{\mathcal{L}}$,

$$\mathcal{W}(x)=\bigcap\{\bigcup\{\mathcal{V}'(x)\colon \mathcal{M}' \preccurlyeq^x_R \mathcal{M}''\}\colon \mathcal{M}'' \succcurlyeq^x_R \mathcal{M}\}.$$

Intuitively, this corresponds to a valuational policy where the things to be done in order to evaluate an expression x are reversed: here one must first fill in the gaps (i.e., look at the (x,R)-completions of the given model) and *then* see whether there is any way of weeding out the gluts (looking at the (x,R)-constrictions) which agrees on some common value. In particular, if R is the universal relation holding among all models of $\mathcal{L}$, it is easy to verify that the R-valuation determined by any $\mathcal{M}\in MOD_{\mathcal{L}}$ is precisely the valuation discussed above:

$$\mathcal{V}(x)=\bigcup\{\bigcap\{\mathcal{V}'(x)\colon \mathcal{M}'\succcurlyeq\mathcal{M}''\}\colon \mathcal{M}''\preccurlyeq\mathcal{M}\}$$

where each $\mathcal{V}'$ is a full homomorphism. (That is, if $R=MOD_{\mathcal{L}}\times MOD_{\mathcal{L}}$, then reference to (x,R)-constrictions and (x,R)-completions yields the same outcome as reference to constrictions and completions *tout court*.) In this case the corresponding dual is defined by setting:

$$\mathcal{W}(x)=\bigcap\{\bigcup\{\mathcal{V}'(x)\colon \mathcal{M}'\preccurlyeq\mathcal{M}''\}\colon \mathcal{M}''\succcurlyeq\mathcal{M}\};$$

and it would be natural to expect that $\mathcal{W}=\mathcal{V}$. For along with

$$\mathcal{M}=\bigsqcup\{\bigsqcap\{\mathcal{M}'\colon \mathcal{M}'\succcurlyeq\mathcal{M}''\}\colon\mathcal{M}''\preccurlyeq\mathcal{M}\},$$

the dual identity

$$\mathcal{M}=\bigsqcap\{\bigsqcup\{\mathcal{M}'\colon \mathcal{M}'\preccurlyeq\mathcal{M}''\}\colon \mathcal{M}''\succcurlyeq\mathcal{M}\}$$

is also satisfied by every model (of any given language). Rather surprisingly, however, this is not the case. That is, the identity $\mathcal{V}=\mathcal{W}$ may fail. Suppose, for example, that $\mathcal{L}$ is a sentential language with a binary connective $\downarrow$, as in Example 1.1.4(A), and suppose $\mathcal{M}$ is a 2-valued model for $\mathcal{L}$ interpreting $\downarrow$ as the joint-denial connective, as in Example 1.2.4(A). Pick $c\in C_{\mathcal{M}}$. If $p=\mathbf{s}(\zeta,0)$ and $q=\mathbf{s}(\xi,0)$ are sentence symbols such that $\mathbf{d}(\zeta,0)[c]=\{0,1\}$ and $\mathbf{d}(\xi,0)[c]=\varnothing$, then it is easy to verify that the original definition yields a gap $\mathcal{V}(p\leftrightarrow q)[c]=\varnothing$ whereas its dual variant would yield a glut $\mathcal{W}(p\leftrightarrow q)[c]=\{0,1\}$.[39] More generally, it can be shown for every language $\mathcal{L}$ and every $\mathcal{M}\in MOD_{\mathcal{L}}$ that the value assignments of the original valuation $\mathcal{V}$ are always included in those of the corresponding dual valuation $\mathcal{W}$ (since every completion of any constriction of a model $\mathcal{M}$ is extended by some constriction of some completion of $\mathcal{M}$ itself), whereas the converse may fail unless $\mathcal{M}$ is ei-

---

[39] The example is based on a remark of Visser [1984], p. 194.

ther complete or consistent (for in such cases $\mathcal{V}(x)=\mathcal{W}(x)$ reduces to $\bigcup\{\mathcal{V}'(x)\colon \mathcal{M}'\preccurlyeq\mathcal{M}\}$ and to $\bigcap\{\mathcal{V}'\colon(x)\ \mathcal{M}'\succcurlyeq\mathcal{M}\}$, respectively).

Now this fact has some intrinsic interest, but it must be properly assessed especially inasmuch as it affects our initial arguments in support of definition 1.3.2. A major justification for this definition was its *direct* bearing upon the possibility of representing any model by a class of sharp models, as indicated in (6'). More generally, 1.3.2 makes intuitive sense when R is an ideal admissibility relation. Yet, to the extent that an alternative account exploiting an equivalent fact is also available, the question arises of *what* criteria could possibly justify a preference for a $\cup\cap$-valuation $\mathcal{V}$ as opposed to its $\cap\cup$-dual $\mathcal{W}$. And this seems to mark an *impasse* in our concern to avoid battles of intuitions at the foundational level.

There is no immediate way out of this apparent dilemma. Nonetheless, I think there is a good explanation for it within our approach. For in the final analysis this situation ties in directly with the assumption that inconsistency and incompleteness are to be treated *on a par*. Certainly that is a most natural and neutral view, given the intrinsic duality of these two notions. Yet it should come as no surprise that when gluts and gaps occur simultaneously, our approach forces us to choose among two possible policies: we can either weed out the gluts first, or we can start by filling in the gaps instead. And although these two policies yield the same result in most cases, in those cases when the very expression to be evaluated amounts to a direct *comparison* between gluts and gaps (as in the above example) the difference must become apparent: on one policy gluts go first, hence gaps prevail; on the other policy it is gaps that go first and gluts prevail. (This point becomes clearer if one considers that, in the case of a 2-valued sentential model of the sort mentioned above, a biconditional $(p\leftrightarrow q)$ turns out to be the only type of sentence—up to equivalence—on which the two policies may disagree. Similarly, if $\mathcal{L}$ is an elementary language and $\mathcal{M}$ a 2-valued, logically adequate model interpreting a binary predicate $\equiv$ as identity, it can be shown that a sentence of the form $(a\equiv b)$[40] is the only type of sentence, along with biconditionals, where the two valuations may yield different outcomes.)

---

[40] Here and below I write '$(a\equiv b)$' as short for '$\mathbf{g}(\mathbf{g}(\equiv,a),b)$'.

From this point of view, the existence of two equally reasonable valuational policies is not a sign of weakness in the approach. Rather, it reflects an important feature thereof: somehow each model *is* associated with both types of valuations. In this sense, the preference for 1.3.2 over its dual variant may very well be interpreted as a bias toward gaps over gluts. But as we shall see, not much of what follows depends on this particular stand. Most of our results and remarks would continue to hold even if definition 1.3.2 were expanded so as to include the dual of any valuation as well.

Note incidentally that the $\cap\cup$-valuation $\mathcal{W}$ is not the only alternative to definition 1.3.2. One could say that, relative to any incomplete and inconsistent model $\mathcal{M}$ of a given language $\mathcal{L}$, there is a whole family of equally legitimate (but increasingly stronger) potential valuations of $\mathcal{L}$ on $\mathcal{M}$—a family forming a complete lattice under $\sqsubseteq$, with $\mathcal{V}$ at the bottom and $\mathcal{W}$ at the top. However, the intermediate elements of this lattice need not be very "natural", as they would somewhat arbitrarily shift from $\mathcal{V}$ to $\mathcal{W}$ for different expressions. I shall accordingly refrain from considering such cases in any detail. For the same reason, I shall not examine other valuational policies that might at this point come to mind, such as the ones reflected in the following equations:

$$\mathcal{W}_1(x)=\bigcap\{\bigcap\{\mathcal{V}'(x): \mathcal{M}'\succcurlyeq^x_R\mathcal{M}''\}: \mathcal{M}''\preccurlyeq^x_R\mathcal{M}\}$$
$$\mathcal{W}_2(x)=\bigcap\{\bigcap\{\mathcal{V}'(x): \mathcal{M}'\preccurlyeq^x_R\mathcal{M}''\}: \mathcal{M}''\succcurlyeq^x_R\mathcal{M}\}$$
$$\mathcal{W}_3(x)=\bigcup\{\bigcup\{\mathcal{V}'(x): \mathcal{M}'\succcurlyeq^x_R\mathcal{M}''\}: \mathcal{M}''\preccurlyeq^x_R\mathcal{M}\}$$
$$\mathcal{W}_4(x)=\bigcup\{\bigcup\{\mathcal{V}'(x): \mathcal{M}'\preccurlyeq^x_R\mathcal{M}''\}: \mathcal{M}''\succcurlyeq^x_R\mathcal{M}\}.$$

More generally, one could consider any pair of operations f_1 and f_2 in place of $\bigcap$ and $\bigcup$, or even any ordered sequence $F=\langle f_0,\ldots,f_n\rangle$, where each f_i may be thought of as having a relevant (ideal) admissibility relation R_i associated with it. The valuations above as well as the dual valuations $\mathcal{V}$ and $\mathcal{W}$ would then be instances of the general scheme:

$$\mathcal{W}_{\mathrm{F}}(x)=f_0\{\ldots\{f_n\{\mathcal{V}_n(x): \mathcal{M}_nS_n\mathcal{M}_{n-1}\}:\ldots\}: \mathcal{M}_0S_0\mathcal{M}\},$$

where each S_i is either $\succcurlyeq^x_{R_i}$ or $\preccurlyeq^x_{R_i}$. It would be interesting to pursue abstract studies in the properties of this general scheme. However, that is not my purpose here. Although there might be some intuitive content to some of these alternatives to definition 1.3.2, there is no connection between them and the general structural properties of $\mathbb{M}$ expressed in

(6').[41] By contrast, it is precisely because of such a connection that $\mathcal{W}$ could be regarded as a serious alternative to $\mathcal{V}$.

At this point the main features of our general semantic framework should be clear. I shall now provide some concrete examples of the behavior of definition 1.3.2 in connection with the types of languages already considered in some detail earlier on.

1.3.4. EXAMPLES

Let $\mathcal{L}$ be a language, $\mathcal{M}$ a model for $\mathcal{L}$, and $\mathcal{V}$ the valuation of $\mathcal{L}$ on $\mathcal{M}$ determined by an ideal admissibility relation R. Then:

(A) If $\mathcal{L}$ is a sentential language with a binary connective $\mathbf{s}(\beta,t)=\downarrow$, as in Example 1.1.4(A), $\mathcal{M}$ is a 2-valued, extensionally adequate model for $\mathcal{L}$, as in Example 1.2.4(A), and $R\subseteq K\times K$, where K is the class of all such models, then the following conditions hold for all $A,B\in\mathbf{E}_0$ and all $c\in C_{\mathcal{M}}$:

(a) $0\in\mathcal{V}_c(\neg A)$ if and only if $1\in\mathcal{V}_c(A)$
$1\in\mathcal{V}_c(\neg A)$ if and only if $0\in\mathcal{V}_c(A)$
(b) $0\in\mathcal{V}_c(A\downarrow B)$ if $1\in\mathcal{V}_c(A)$ or $1\in\mathcal{V}_c(B)$
$1\in\mathcal{V}_c(A\downarrow B)$ only if $0\in\mathcal{V}_c(A)$ and $0\in\mathcal{V}_c(B)$
(c) $0\in\mathcal{V}_c(A\vee B)$ only if $0\in\mathcal{V}_c(A)$ and $0\in\mathcal{V}_c(B)$
$1\in\mathcal{V}_c(A\vee B)$ if $1\in\mathcal{V}_c(A)$ or $1\in\mathcal{V}_c(B)$
(d) $0\in\mathcal{V}_c(A\wedge B)$ if $0\in\mathcal{V}_c(A)$ or $0\in\mathcal{V}_c(B)$
$1\in\mathcal{V}_c(A\wedge B)$ only if $1\in\mathcal{V}_c(A)$ and $1\in\mathcal{V}_c(B)$
(e) $0\in\mathcal{V}_c(A\rightarrow B)$ only if $1\in\mathcal{V}_c(A)$ and $0\in\mathcal{V}_c(B)$
$1\in\mathcal{V}_c(A\rightarrow B)$ if $0\in\mathcal{V}_c(A)$ or $1\in\mathcal{V}_c(B)$

[*Proof.* I only consider (b), since the other claims follow directly by 1.1.4(A)(a)–(d). As a lemma, suppose first that $\mathcal{M}\in SHRP_{\mathcal{L}}(A\downarrow B)$. In that case (b) follows directly by 1.2.4(A)(*) via the identities $\mathcal{V}(A\downarrow B)=\mathcal{V}(\mathbf{g}(\mathbf{g}(\downarrow,A),B))=\mathbf{h}(\mathcal{V}(\mathbf{g}(\downarrow,A),\mathcal{V}(B))=\mathbf{h}(\mathbf{h}(\mathcal{V}(\downarrow),\mathcal{V}(A)),\mathcal{V}(B))=\mathbf{h}(\mathbf{h}(\mathbf{d}(\beta,t),\mathcal{V}(A)),\mathcal{V}(B))$. (This implies that R-valuations on models in

[41] I have studied some of these possibilities in Varzi [1994c]. A valuational strategy corresponding to the schema $\mathcal{W}_1$ is proposed in Lewis [1978], though Lewis [1983b] suggests amending it along the lines of a ∪∩-valuation.

$SHRP_{\mathcal{L}}(A{\downarrow}B)$ satisfy also the converses of both clauses in (b), hence of all clauses in (c)–(e) [42]). Suppose now that $\mathcal{M}\notin SHRP_{\mathcal{L}}(A{\downarrow}B)$, and consider the first clause of (b). If $1\in\mathcal{V}_c(A)\cup\mathcal{V}_c(B)$, then there exist $X\in\{A,B\}$ and $\mathcal{M}''\preccurlyeq^X_R\mathcal{M}$ such that $1\in\mathcal{V}''_c(X)=\bigcap\{\mathcal{V}'_c(X)\colon \mathcal{M}'\succcurlyeq^X_R\mathcal{M}''\}$. Let $\mathcal{M}^*\sqsubseteq\mathcal{M}''$ be $\sqsubseteq$-maximal in $K\cap CONS_{\mathcal{L}}(A{\downarrow}B)$. Then $\mathcal{M}^*\preccurlyeq^{A\downarrow B}_R\mathcal{M}$, and by maximality we have $1\in\mathcal{V}^*_c(X)=\bigcap\{\mathcal{V}'_c(X)\colon \mathcal{M}'\succcurlyeq^X_R\mathcal{M}^*\}$. Hence $1\in\bigcap\{\mathcal{V}'_c(X)\colon \mathcal{M}'\succcurlyeq^{A\downarrow B}_R\mathcal{M}^*\}$ (since every $(A{\downarrow}B,R)$-completion is also an (X,R)-completion). Therefore $0\in\bigcap\{\mathcal{V}'_c(A{\downarrow}B)\colon \mathcal{M}'\succcurlyeq^{A\downarrow B}_R\mathcal{M}^*\}$ (by the lemma), which implies $0\in\mathcal{V}_c(A{\downarrow}B)$. Consider now the second clause. If $1\in\mathcal{V}_c(A{\downarrow}B)$, then there is some $\mathcal{M}^*\preccurlyeq^{A\downarrow B}_R\mathcal{M}$ so that $1\in\mathcal{V}^*_c(A{\downarrow}B)=\bigcap\{\mathcal{V}'_c(A{\downarrow}B)\colon \mathcal{M}'\succcurlyeq^{A\downarrow B}_R\mathcal{M}^*\}$, and thus $0\in\bigcap\{\mathcal{V}'_c(X)\colon \mathcal{M}'\succcurlyeq^{A\downarrow B}_R\mathcal{M}^*\}$ for both $X=A$ and $X=B$ (by the lemma). Since any such $\mathcal{M}^*$ is also an (X,R)-constriction of $\mathcal{M}$, it follows that $0\in\mathcal{V}_c(A)$ and $0\in\mathcal{V}_c(B)$.]

As a concrete case study for these properties, suppose for simplicity that $IND_{\mathcal{L}}$ contains only two sentence symbols, $\mathbf{s}(\beta,0)=p$ and $\mathbf{s}(\gamma,0)=q$. If we only focus upon the class of all 2-valued, extensionally adequate sentential models for $\mathcal{L}$ with a single context c, then the overall picture is as in Figure 1. With reference to this picture it is easy to verify (a)–(e) as well as to find counterexamples to the converse implications. For example, if $\mathcal{M}$ is an incomplete model in the right portion of the diagram, with $\mathbf{d}(\beta,0)[c]=\varnothing$, then $0\in\mathcal{V}_c(p\wedge\neg p)$ and $1\in\mathcal{V}_c(p\vee\neg p)$ even though $0,1\notin\mathcal{V}_c(p)=\mathcal{V}_c(\neg p)$, while if $\mathcal{M}$ is an inconsistent model in the left portion of the diagram, with $\mathbf{d}(\beta,0)[c]=\{0,1\}$, then $1\notin\mathcal{V}_c(p\wedge\neg p)$ and $0\notin\mathcal{V}_c(p\vee\neg p)$ even though $1,0\in\mathcal{V}_c(p)=\mathcal{V}_c(\neg p)$. Note also that if $\mathcal{M}$ is sharp, then for every $A,B\in\mathbf{I}_0$ we have $0\in\mathcal{V}_c(A\leftrightarrow B)$ if and only if $\mathcal{V}_c(A)\neq\mathcal{V}_c(B)$, and $1\in\mathcal{V}_c(A\leftrightarrow B)$ if and only if $\mathcal{V}_c(A)=\mathcal{V}_c(B)$. These standard conditions are preserved as implications (from left to right and from right to left, respectively) in the case of models that are either consistent or complete, but no general condition can be specified for the two incomplete and inconsistent models on the side vertices of the diagram. In particular, notice that such models yield the gaps $\mathcal{V}_c(p\leftrightarrow q)=\mathcal{V}_c(q\leftrightarrow p)=\varnothing$ (whereas the dual valuational policy mentioned earlier would yield the gluts $\mathcal{W}_c(p\leftrightarrow q)=\mathcal{W}_c(q\leftrightarrow p)=\{0,1\}$).

---

[42] In other words, R-valuations on such models satisfy the relevant conditions of classical 2-valued sentential logic.

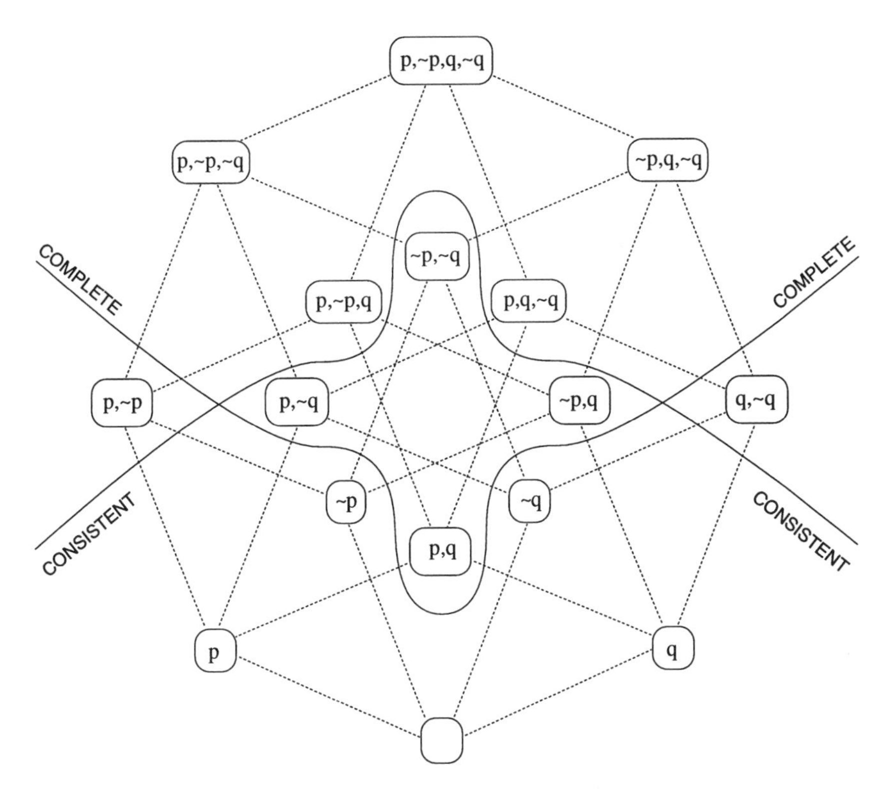

Figure 1. Lattice of 2-valued extensional models (with $C=\{c\}$) for a sentential language $\mathcal{L}$ with only two sentence symbols, p and q. Each node corresponds to a model, with 'p' and '~p' indicating the presence of 1 and 0, respectively, in the denotations of 'p' (in c), and likewise for 'q' and '~q'. The ordering $\sqsubseteq$ goes uphill along the dashed lines.

Incidentally, the above counterexamples to the converse implications in (c) and (d) are not just a consequence of the failure of compositionality. They exhibit a characteristic feature of the present approach, namely, that the requirements of completeness and consistency—in this case, the requirements that every sentence must be *either* true *or* false in c, and no sentence *both* true *and* false, respectively—are not mirrored in the truth conditions of sentences of the form '$p \vee \neg p$' and '$p \wedge \neg p$', even if these conditions are computed on a 2-valued sentential model. The latter are sentences of $\mathcal{L}$, the former semantic requirements. And while the distinction turns out to be empty in classical contexts, where both semantic requirements are satisfied, the acknowledgement of gaps and gluts gives it content: completeness and consistency may fail in some models, but the corresponding disjunction and conjunction retain their ordinary truth conditions.[43] Thus, intuitively, in the world described by the Holmes stories both 'Watson limps' and 'Watson does not limp' are true (and both are false) because some stories contradict each other regarding the location of Watson's old war wound. But no story involves the contradiction 'Watson both does and does not limp', which therefore will not be true. In fact it can only be evaluated as false, since it is false in each story. Likewise, neither 'Watson was born on a Tuesday' nor 'Watson was not born on a Tuesday' is true (or false) if the stories say nothing about Watson's birth date. Nevertheless the disjunction 'Watson either was or was not born on a Tuesday' will come out true (and only true) since this truth-value is independent on how we tell the stories.

(B) If $\mathcal{L}$ is an elementary language with a set $V_{\mathcal{L}}$ of name variables and a binary quantifier $\mathbf{d}(\beta,t)=\langle\downarrow,v\rangle$ for each $v\in V_{\mathcal{L}}$, as in Example 1.1.4 (B), $\mathcal{M}$ is a 2-valued, non-empty, logically adequate elementary model for $\mathcal{L}$, as in Example 1.2.4(B), and $R\subseteq K\times K$, where K is the class of such models, then conditions (a)–(e) above along with the following conditions hold for all $A,B\in\mathbf{E}_0$, all $v\in V_{\mathcal{L}}$, and all $c\in C_{\mathcal{M}}=\mathbf{D}_1{}^{V_{\mathcal{L}}}$:

(f) $0\in\mathcal{V}_c(A\downarrow^v B)$ if $1\in\mathcal{V}_{c(v/u)}(A)$ or $1\in\mathcal{V}_{c(v/u)}(B)$ for some $u\in\mathbf{D}_1$
$1\in\mathcal{V}_c(A\downarrow^v B)$ only if $0\in\mathcal{V}_{c(v/u)}(A)$ and $0\in\mathcal{V}_{c(v/u)}(B)$ for all $u\in\mathbf{D}_1$
(g) $0\in\mathcal{V}_c(\bigvee vA)$ only if $0\in\mathcal{V}_{c(v/u)}(A)$ for all $u\in\mathbf{D}_1$
$1\in\mathcal{V}_c(\bigvee vA)$ if $1\in\mathcal{V}_{c(v/u)}(A)$ for some $u\in\mathbf{D}_1$
(h) $0\in\mathcal{V}_c(\bigwedge vA)$ if $0\in\mathcal{V}_{c(v/u)}(A)$ for some $u\in\mathbf{D}_1$
$1\in\mathcal{V}_c(\bigwedge vA)$ only if $1\in\mathcal{V}_{c(v/u)}(A)$ for all $u\in\mathbf{D}_1$

[*Proof.* Note that the assumption of non-emptiness is needed to ensure the possibility that R be ideal. Otherwise the definition of K would imply that $\mathbf{D}_1'=\varnothing$ for every admissible extension of $\mathcal{M}$. This being said, we only need to consider (f), whence the rest follows immediately by 1.1.4(B)(a)–(g). The argument parallels the one of Example (A). As a lemma, note in this case that if $\mathcal{M}\in SHRP_{\mathcal{L}}(A\downarrow^v B)$, then both clauses in (f) along with their converses follow from 1.2.4(B)(c) via the basic identities $\mathcal{V}(A\downarrow^v B)=\mathcal{V}(\mathbf{g}(\mathbf{g}(\langle\downarrow,v\rangle,A),B))=\mathbf{h}(\mathcal{V}(\mathbf{g}(\langle\downarrow,v\rangle,A),\mathcal{V}(B))=$

[43] Most of the works listed in note 34 share this feature. See also below, note 52 and section 2.1.4(A), where this point is examined in greater detail.

$\mathbf{h}(\mathbf{h}(\mathcal{V}(\langle\downarrow,v\rangle),\mathcal{V}(A)),\mathcal{V}(B))=\mathbf{h}(\mathbf{h}(\mathbf{d}(\beta,t),\mathcal{V}(A)),\mathcal{V}(B))$. Suppose now that $\mathcal{M}\notin SHRP_{\mathcal{L}}(A\downarrow^{v}B)$ and consider the first clause in (f). If there exists some $u\in\mathbf{D}_1$ such that $1\in\mathcal{V}_{c(v/u)}(A)\cup\mathcal{V}_{c(v/u)}(B)$, then $1\in\mathcal{V}''_{c(v/u)}(X)=\bigcap\{\mathcal{V}'_{c(v/u)}(X)\colon \mathcal{M}'\succcurlyeq^{X}_{R}\mathcal{M}''\}$ for some $X\in\{A,B\}$ and some $\mathcal{M}''\preccurlyeq^{X}_{R}\mathcal{M}$. Let $\mathcal{M}^*\sqsubseteq\mathcal{M}''$ be $\sqsubseteq$-maximal in $K\cap CONS_{\mathcal{L}}(A\downarrow^{v}B)$. Then $\mathcal{M}^*\preccurlyeq^{A\downarrow^{v}B}_{R}\mathcal{M}$, and consequently $1\in\mathcal{V}^*_{c(v/u)}(X)=\bigcap\{\mathcal{V}'_{c(v/u)}(X)\colon \mathcal{M}'\succcurlyeq^{X}_{R}\mathcal{M}^*\}$ by maximality. Hence $1\in\bigcap\{\mathcal{V}'_{c(v/u)}(X)\colon \mathcal{M}'\succcurlyeq^{A\downarrow^{v}B}_{R}\mathcal{M}^*\}$ (since every $(A\downarrow^{v}B,R)$-completion is also an (X,R)-completion). But $\mathbf{D}_1\subseteq\mathbf{D}'_1$. Therefore we conclude that $0\in\bigcap\{\mathcal{V}'_{c(v/u)}(A\downarrow^{v}B)\colon \mathcal{M}'\succcurlyeq^{A\downarrow^{v}B}_{R}\mathcal{M}^*\}$ (by the lemma), which implies that $0\in\mathcal{V}_c(A\downarrow^{v}B)$. The proof of the second clause is similar, as already seen in example (A). As a corollary, note that if the applicable sharpness conditions are fulfilled, the converses of the clauses in (g)–(h) hold too.[44] Also, we then have $1\in\mathcal{V}_c(A\leftrightarrow B)$ if and only if $\mathcal{V}_c(A)=\mathcal{V}_c(B)$, and $0\in\mathcal{V}_c(A\leftrightarrow B)$ if and only if $\mathcal{V}_c(A)\neq\mathcal{V}_c(B)$. As in example (A), however, no such properties can be specified for $\mathcal{V}_c(A\leftrightarrow B)$ in the general case, where $\mathcal{M}$ may be both incomplete and inconsistent.]

To illustrate, suppose that $IND_{\mathcal{L}}-V_{\mathcal{L}}$ contains only two name symbols, say $\mathbf{s}(\beta,1)=a$ and $\mathbf{s}(\gamma,1)=b$, and consider a 2-valued, logically adequate model $\mathcal{M}$ whose domain $\mathbf{D}_1$ contains exactly two objects, say u and w. Suppose, further, that $\mathcal{M}$ involves exactly two gaps and two gluts in each context (i.e., value assignment). For instance, suppose that, for each $c\in C_{\mathcal{M}}$, $\mathbf{d}(\beta,1)[c]=\varnothing$ and $\mathbf{d}(\gamma,1)[c]=\{u,w\}$ while $\mathbf{h}(x,y)[c]=\varnothing$ and $\mathbf{h}(x',y')[c]=\{0,1\}$, where $x,x'\in\mathbf{I}_{\langle 1,0\rangle}$ and $y,y'\in\mathbf{I}_1$ are constant functions. Finally, suppose for simplicity that $R=K\times K$. In that case, (x,R)-constrictions/completions can be dispensed with in favor of full constrictions/completions (as remarked in section 1.3.1). Hence the portion of $MOD_{\mathcal{L}}$ containing $\mathcal{M}$ along with the relevant sharpenings can be depicted as in Figure 2. Though partial, this picture is sufficient to verify clauses (f)–(h) and to find counterexamples to the converse implications. Consider (g), for instance, and suppose $FUN_{\mathcal{L}}$ contains a binary predicate $\mathbf{s}(\delta,\langle 1,\langle 1,0\rangle\rangle)=\equiv$ interpreted in $\mathcal{M}$ as singular identity: for all $x,y\in\mathbf{I}_1$ and all $c\in C_{\mathcal{M}}$, $\mathbf{h}(\mathbf{h}(\mathbf{d}(\delta,\langle 1,\langle 1,0\rangle\rangle),x),y)(c)=1$ if $x(c)=y(c)$, and $\mathbf{h}(\mathbf{h}(\mathbf{d}(\delta,\langle 1,\langle 1,0\rangle\rangle),x),y)(c)=0$ otherwise. Then, for each $c\in C_{\mathcal{M}}$ we have $1\in\mathcal{V}_c(\bigvee v(v\equiv a))$ even if $1\notin\mathcal{V}_{c(v/u)}(v\equiv a)$ for all $u\in\mathbf{D}_1$, and conversely,

---

[44] That is, one gets the conditions of classical 2-valued elementary logic.

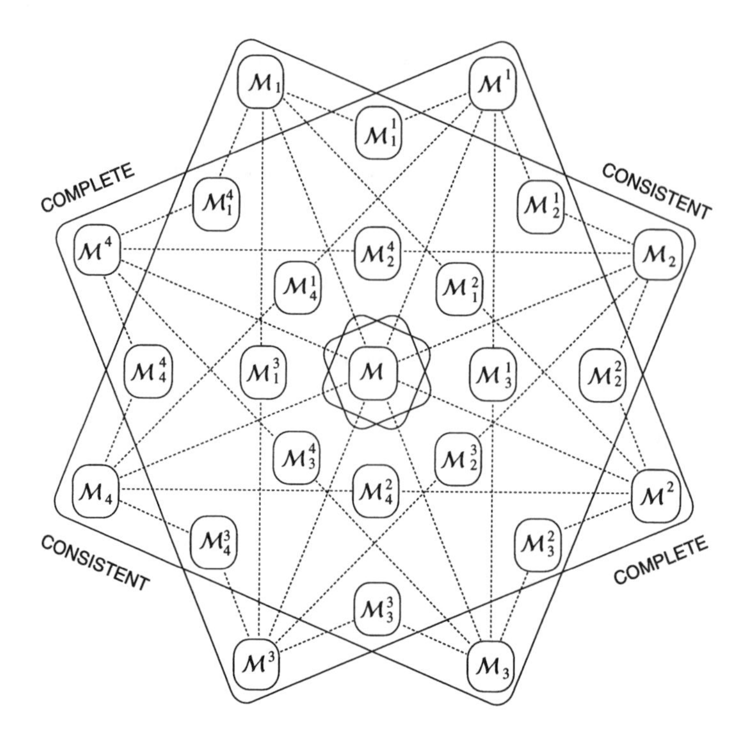

Figure 2. Lattice of (full) sharpenings of a 2-valued, logically adequate elementary model $\mathcal{M}$ with two gaps and two gluts. The picture is seen from above, as it were: at the bottom are the four possible constrictions of $\mathcal{M}$ ($\mathcal{M}_1$, $\mathcal{M}_2$, $\mathcal{M}_3$, and $\mathcal{M}_4$); at the top are the four completions ($\mathcal{M}^1$, $\mathcal{M}^2$, $\mathcal{M}^3$, and $\mathcal{M}^4$); all other models in the picture lie at an intermediate level and comprise $\mathcal{M}$ itself plus the sixteen logically adequate sharp models obtained by weeding out the gluts in $\mathcal{M}$'s completions or, equivalently, by filling in the gaps in $\mathcal{M}$'s constrictions: for instance, $\mathcal{M}_j^i$ is a completion of $\mathcal{M}_j$, but also a constriction of $\mathcal{M}^i$ ($1 \leq i,j \leq 4$).

$0 \notin \mathcal{V}_c(\bigvee v(v \equiv b))$ even if $0 \in \mathcal{V}_{c(v/u)}(v \equiv b)$ for all $u \in \mathbf{D}_1$. The other cases are similarly checked. Moreover, it is now easy to verify that $\mathcal{V}_c(a \equiv b) = \mathcal{V}_c(b \equiv a) = \varnothing$ for all $c \in C_{\mathcal{M}}$, while the dual valuational policy would yield $\mathcal{W}_c(a \equiv b) = \mathcal{W}_c(b \equiv a) = \{0,1\}$. This difference reproduces at the predicate level the pattern of disagreement between $\mathcal{V}$ and $\mathcal{W}$ featured at the sentential level by biconditionals (as in the previous example).

Also in this case, it is worth noting that our counterexamples to the converse implications in (b)–(g) are not mere consequence of the failure of compositionality. They exhibit, rather, a characteristic feature of the present approach already encountered in example (A). For instance, just as a disjunction $A \vee \neg A$ may not be used to express completeness with respect to a sentence A, so a quantified formula $\bigvee v(v \equiv a)$ may not be used to express completeness with respect to a name a: For just as $A \vee \neg A$ can be true even when A is truth-valueless, in virtue of a truth that shifts from one disjunct to the other for different ways of evaluating A, so $\bigvee v(v \equiv a)$ can be true (under an assignment c) even when a is denotationless (under that assignment) in virtue of a truth that shifts from one instance to another for different ways of interpreting a.[45] Thus, for instance, in the world of the Holmes stories the sentence 'Some date or other is Watson's birth date' is true, even though its substitution instances—sentences such as 'Watson's birth date is December 6, 1850'—are all truth-valueless.

(C) If $\mathcal{L}$ is a full categorial language with a denumerable set $V_{\mathcal{L},t}$ of variables for each type $t \in \mathcal{T}$ and an abstractor $\langle \lambda, t', v \rangle$ for all $v \in V_{\mathcal{L},t}$ and all $t' \in \mathcal{T}$, as in Example 1.1.4(C), if $\mathcal{M}$ is a 2-valued, combinatorially adequate model for $\mathcal{L}$, as in Example 1.2.4(C), and if $R = K \times K$, where K is the class of such models, then the following conditions hold for all $v \in \bigcup\{V_{\mathcal{L},t}: t \in \mathcal{T}\}$, all $x \in \mathbf{E}_{\tau(v)}$, all $t, t', t'' \in \mathcal{T}$, and all $A \in \mathbf{E}_{t'}$, $B \in \mathbf{E}_t$, $C \in \mathbf{E}_{\langle t,t' \rangle}$, and $D \in \mathbf{E}_{\langle t, \langle t', t'' \rangle \rangle}$:

(a) $\mathcal{V}((\lambda v A)x) = \mathcal{V}(A^{x}/_{v})$
(b) $\mathcal{V}(\mathbf{I}_t(B)) = \mathcal{V}(B)$
(c) $\mathcal{V}(\mathbf{K}_{t,t'}(B,A)) = \mathcal{V}(B)$
(d) $\mathcal{V}(\mathbf{S}_{t,t',t''}(D,C,B)) = \mathcal{V}(D(B,C(B)))$

where $A^{x}/_{v}$—the result of substituting x for v in A—is defined as follows: (i) if $A \in SYM_{\mathcal{L}}$, then $A^{x}/_{v} = x$ if $A = v$ and $A^{x}/_{v} = A$ otherwise; (ii) if $A = \mathbf{g}(x,y)$, then $A^{x}/_{v} = \mathbf{g}(y^{x}/_{v}, z^{x}/_{v})$.

[*Proof*. Here note that if $\mathcal{M} \in SHRP_{\mathcal{L}}((\lambda v A)x)$ and $\mathbf{s}(\beta,t) = \langle \lambda, t', v \rangle$, then $\mathcal{V}((\lambda v A)x) = \mathbf{h}(\mathbf{h}(\mathbf{d}(\beta,t), \mathcal{V}(A)), \mathcal{V}(x))$, and thus $\mathcal{V}((\lambda v A)x)(c) = \mathcal{V}(A)(c(\tau(v)/c_{\tau(v)}(v/x(c))))$ for all $c \in C_{\mathcal{M}} = \prod \langle \mathbf{D}_t^{V_{\mathcal{L},t}}: t \in \mathcal{T} \rangle$ by 1.2.4(C)(c). Moreover, an inductive argument shows that if $\mathcal{M} \in SHRP_{\mathcal{L}}(A^{x}/_{v})$, then the "substitution lemma" holds, i.e., $\mathcal{V}(A^{x}/_{v})(c) = \mathcal{V}(A)(c(\tau(v)/c_{\tau(v)}(v/x(c))))$

---

[45] From this point of view, the present example parallels the semantics of Fine [1975]. See below, note 59 and § 2.1.4(B), for more details on this point.

for all $c \in C_{\mathcal{M}}$. All this implies that (a) holds whenever the relevant sharpness conditions are fulfilled. But then it follows that $(\lambda v A)x$ and $A^{x}/_{v}$ will receive the same value also when those conditions are not fulfilled, for each value assignment is determined by the relevant sharpenings, and the condition $R = K \times K$ guarantees that all expressions have the same sharpenings. Given (a), (b)–(d) then follow easily by definition.[46]]

1.3.5. AUXILIARY NOTIONS

Some important semantic notions, such as the notions of validity, entailment, and derivatives, can easily be generalized within the present framework. For this purpose, where $\mathcal{L}$ is any language, let us define the auxiliary concept of a *canonical frame* for $\mathcal{L}$ (of type t) to be a fourtuple

$$\mathcal{F} = \langle K, R, t, \delta \rangle$$

where $K \subseteq MOD_{\mathcal{L}}$, $R \subseteq K \times K$, $t \in \mathcal{DE}$, and δ is a K-termed sequence $\langle \delta_{\mathcal{M}} \in \mathbf{D}_t \colon \mathcal{M} \in K \rangle$ with the property that $\delta_{\mathcal{M}} = \delta_{\mathcal{M}'}$ whenever $\mathcal{M} R \mathcal{M}'$. Intuitively, K may be thought of as a class of "admissible" models, R as the corresponding admissibility relation, and each $\delta_{\mathcal{M}}$ as specifying a "designated value" of type t—for instance, a designated truth-value if $t = 0$, a designated individual if $t = 1$, and so on. (The requirement on $\delta_{\mathcal{M}} = \delta_{\mathcal{M}'}$ is to guarantee certain intuitive conditions of stability. In some cases, e.g., when $t = 0$, it is natural to go further and require δ to pick out the same value in every admissible model.)

Canonical frames define a frame of reference, as it were. Given $\mathcal{F} = \langle K, R, t, \delta \rangle$ we can say, for instance, that an expression $x \in \mathbf{E}_t$ is satisfied in a model $\mathcal{M} \in K$ by a context $c \in C_{\mathcal{M}}$ if the values of x in c under R include the designated value of $\mathcal{F}$. More precisely:

(a) x is *$\mathcal{F}$-satisfied* in $\mathcal{M}$ by c (written $\mathcal{M} \Vdash_{\mathcal{F}}^{c} x$) iff $x \in \mathbf{E}_t$, $\mathcal{M} \in K$, $c \in C_{\mathcal{M}}$, and $\delta_{\mathcal{M}} \in \mathcal{V}_{R,c}(x)$.

Then we can define, for all $x, y \in EXP_{\mathcal{L}}$:

---

[46] These conditions reflect some standard properties of categorial languages with combinators.

(b) x is *$\mathcal{F}$-valid* iff $\mathcal{M} \Vdash_{\mathcal{F}}^{c} x$ for all $\mathcal{M} \in K$ and all $c \in C_{\mathcal{M}}$
(c) x is *$\mathcal{F}$-entailed* by y iff $\mathcal{M} \Vdash_{\mathcal{F}}^{c} x$ for all $\mathcal{M} \in K$ and all $c \in C_{\mathcal{M}}$ such that $\mathcal{M} \Vdash_{\mathcal{F}}^{c} y$.

More generally, let us associate every $x \in \mathsf{EXP}_{\mathcal{L}}$ with its *$\mathcal{F}$-satisfaction class*:

$$\mathsf{SAT}_{\mathcal{F}}(x) = \{\langle \mathcal{M}, c \rangle : \mathcal{M} \Vdash_{\mathcal{F}}^{c} x\}$$

and let us agree in the present context to identify $\bigcap\varnothing$ with the class

$$\mathcal{U}_{\mathcal{F}} = \{\langle \mathcal{M}, c \rangle : \mathcal{M} \in K \text{ and } c \in C_{\mathcal{M}}\}.$$

Then we can define, for all sets of expressions $X, Y \subseteq \mathsf{EXP}_{\mathcal{L}}$:

(d) X is *$\mathcal{F}$-satisfiable* iff $\bigcap\{\mathsf{SAT}_{\mathcal{F}}(x) : x \in X\} \neq \varnothing$
(e) X is *$\mathcal{F}$-refutable* iff $\bigcup\{\mathsf{SAT}_{\mathcal{F}}(x) : x \in X\} \neq \mathcal{U}_{\mathcal{F}}$
(f) X is *$\mathcal{F}$-entailed* by Y iff $\bigcap\{\mathsf{SAT}_{\mathcal{F}}(y) : y \in Y\} \subseteq \bigcup\{\mathsf{SAT}_{\mathcal{F}}(x) : x \in X\}$.

Various equivalent characterizations of these notions are of course available. For instance, (d)–(f) immediately imply the following:

(d') X is $\mathcal{F}$-satisfiable iff $\varnothing$ is not $\mathcal{F}$-entailed by X
(e') X is $\mathcal{F}$-refutable iff X is not $\mathcal{F}$-entailed by $\varnothing$.

For another example, let us associate each expression $x \in \mathsf{EXP}_{\mathcal{L}}$ with its *$\mathcal{F}$-valuation region*—the set $\mathsf{VAL}_{\mathcal{F}}(x)$ of those valuation fragments (relative to the components of $\mathcal{F}$) which assign x the designated value (possibly along with other values):

$$\mathsf{VAL}_{\mathcal{F}}(x) = \{\mathcal{V}_{R,c} : \mathcal{M} \Vdash_{\mathcal{F}}^{c} x\}.$$

And let us agree in the present context to identify $\bigcap\varnothing$ with the class

$$\mathcal{U}_{R} = \{\mathcal{V}_{R,c} : \mathcal{M} \in K \text{ and } c \in C_{\mathcal{M}}\}.$$

Then (f) can be re-written as:

(f') X is $\mathcal{F}$-entailed by Y iff $\bigcap\{\mathsf{VAL}_{\mathcal{F}}(y) : y \in Y\} \subseteq \bigcup\{\mathsf{VAL}_{\mathcal{F}}(x) : x \in X\}$.

Such equivalences align our generalized notions with their customary counterparts. In particular, let us symbolize the relation of $\mathcal{F}$-entailment by $\vDash_{\mathcal{F}}$. Then one easily determines that this relation enjoys the usual structural properties:

Reflexivity	$X \vDash_{\mathcal{F}} X$
Reiteration	$Y \cup \{z\} \vDash_{\mathcal{F}} \{z\} \cup X$
Thinning	if $Y \vDash_{\mathcal{F}} X$ then $Y \cup \{z\} \vDash_{\mathcal{F}} X$ and $Y \vDash_{\mathcal{F}} \{z\} \cup X$
Cut	if $Y \cup \{z\} \vDash_{\mathcal{F}} X$ and $Y \vDash_{\mathcal{F}} \{z\} \cup X$ then $Y \vDash_{\mathcal{F}} X$.

Indeed, these properties would continue to hold even if one considers frames $\mathcal{F} = \langle K, R, t, \delta \rangle$ which are not "canonical" in the strict sense defined above—for instance, frames where δ is only partially defined on K (e.g., because $\mathbf{D}_t = \varnothing$ for some $\mathcal{M} \in K$), or where the basic admissibility relation R is not included in $K \times K$.

On the other hand, within our generalized framework it is possible to formulate various alternative, non-equivalent accounts of the above notions. For instance, it is rather natural to expand the concept of a canonical frame by adding a fifth coordinate, δ', to be interpreted as specifying an "anti-designated value" $\delta'_{\mathcal{M}} \in \mathbf{D}_t$ for each $\mathcal{M} \in K$. (If $\mathcal{M}$ is a k-valued model and the designated truth-value is k, for instance, the anti-designated truth-value may be 0.) With $SAT'_{\mathcal{F}}$ defined in the obvious way (using δ' in place of δ), we could then consider the following dual of (f) (and similarly for (f')):

(g) X is *negatively $\mathcal{F}$-entailed* by Y ($Y \vDash^{-}_{\mathcal{F}} X$) iff $\bigcap\{SAT'_{\mathcal{F}}(x) \colon x \in X\} \subseteq \bigcup\{SAT'_{\mathcal{F}}(y) \colon y \in Y\}$.

Or we could consider a double-barreled notion, requiring entailment to be preserving with respect to both the designated and the anti-designated values:[47]

(h) X is *doubly $\mathcal{F}$-entailed* by Y ($Y \vDash^{\pm}_{\mathcal{F}} X$) iff $Y \vDash_{\mathcal{F}} X$ and $Y \vDash^{-}_{\mathcal{F}} X$.

More generally still, we could refer to frames of the form $\langle K, R, t, \leqslant \rangle$, where $\leqslant$ is a complete lattice ordering of $\bigcup\{\wp\mathbf{D}_t \colon \mathcal{M} \in K\}$ with the anti-designated value at the bottom and the designated value at the top. The following variant could then be considered:

(i) X is *fully $\mathcal{F}$-entailed* by Y ($Y \vDash^{\leqslant}_{\mathcal{F}} X$) iff, for every $\mathcal{M} \in K$ and every $c \in C_{\mathcal{M}}$, $\bigsqcap_{\leqslant}\{\mathcal{V}_{R,c}(y) \colon y \in Y\} \leqslant \bigsqcup_{\leqslant}\{\mathcal{V}_{R,c}(x) \colon x \in X\}$.

---

[47] This is in the spirit of Blamey [1986], Muskens [1995], and Dunn [1997]; compare also Cleave [1974].

All of these options—and indeed many others—would legitimately qualify as generalizations of the customary notion of entailment, though they need not coincide if gaps and gluts are allowed. For instance, suppose $\mathcal{L}$ is a sentential language of the sort discussed in our examples, with two distinguished sentence symbols $\mathbf{s}(\beta,0)=p$ and $\mathbf{s}(\gamma,0)=q$, and let K be the corresponding class of 2-valued, extensionally adequate models $\mathcal{M}$ such that, for each $c\in C_{\mathcal{M}}$, either $\mathbf{d}(\beta,0)[c]=\mathbf{d}(\gamma,0)[c]$ or $\mathbf{d}(\beta,0)[c]=\varnothing$ and $\mathbf{d}(\gamma,0)[c]=\{0,1\}$. If $\mathcal{F}=\langle K, K\times K, 0, \delta\rangle$, with $\mathcal{R}\delta=\{1\}$ and $\mathcal{R}\delta'=\{0\}$, then we have $q\vDash_{\mathcal{F}}p$ but neither $q\vDash^{-}_{\mathcal{F}}p$ nor $q\vDash^{\pm}_{\mathcal{F}}p$; and we have $q\vDash^{\leqslant}_{\mathcal{F}}p$ only if $\leqslant$ is defined so that $\varnothing\leqslant\{0,1\}$.[48] (To simplify notation, if x and y are any expressions, I shall write $x\vDash_{\mathcal{F}}y$ to abbreviate $\{x\}\vDash_{\mathcal{F}}\{y\}$, and similarly for the other entailment relations.) Also, it is easy to see that $\vDash^{\pm}_{\mathcal{F}}$ would typically satisfy the classical rule of *Contraposition*:

$$Y\vDash^{\pm}_{\mathcal{F}}X \text{ iff } \{\neg x\colon x\in X\}\vDash^{\pm}_{\mathcal{F}}\{\neg y\colon y\in Y\}.$$

This would also be true of $\vDash^{\leqslant}_{\mathcal{F}}$, but not of $\vDash_{\mathcal{F}}$ or $\vDash^{-}_{\mathcal{F}}$: for instance, although $q\vDash_{\mathcal{F}}p$, it is not the case that $\neg p\vDash_{\mathcal{F}}\neg q$.

Likewise, we can define an obvious relation of equivalence between expressions that get the same values in every context of every relevant model:

(j) x is *$\mathcal{F}$-equivalent to y* ($x\cong_{\mathcal{F}}y$) iff $\mathcal{V}_R(x)=\mathcal{V}_R(y)$ for all $\mathcal{M}\in K$.

Evidently, $\cong_{\mathcal{F}}$ is reflexive, symmetric, and transitive—an equivalence relation—and it is easy to check that any two expressions are $\mathcal{F}$-equivalent if and only if they doubly $\mathcal{F}$-entail each other. Yet this is not true if we take plain $\mathcal{F}$-entailment or negative $\mathcal{F}$-entailment instead. For instance, consider a frame $\mathcal{F}$ defined as above, except that now we require that $\mathbf{d}(\beta,0)[c]\neq\mathbf{d}(\gamma,0)[c]$ for all $\mathcal{M}\in K$ and all $c\in C_{\mathcal{M}}$. Suppose, further, that $\mathcal{L}$ includes a third distinguished symbol $\mathbf{s}(\xi,0)=r$ so that $\mathbf{d}(\xi,0)[c]=\{0\}$ for all $\mathcal{M}\in K$ and all $c\in C_{\mathcal{M}}$. Then we have the two $\mathcal{F}$-entailments $r\vDash_{\mathcal{F}}(p\leftrightarrow q)$ and $(p\leftrightarrow q)\vDash_{\mathcal{F}}r$ (since neither r nor $(p\leftrightarrow q)$ ever receives the

---

[48] As in the lattice **L4** introduced in Belnap [1977] (following a suggestion of Dunn [1976]) and extensively discussed in the literature. See in particular the work of Fitting [1989, 1992], Ginsberg [1988, 1990], Thijsse [1992], and Gupta & Belnap [1993].

value 1) but the corresponding $\mathcal{F}$-equivalence fails in those contexts where p is overdetermined and q undetermined (r will get the value 0 whereas $(p \leftrightarrow q)$ will be undetermined—or overdetermined, if we were working with the dual valuation $\mathcal{W}$). Or again, it is easy to see that $\models^{\pm}_{\mathcal{F}}$ satisfies the classical law of *Correspondence*:

$$x \models^{\pm}_{\mathcal{F}} y \text{ iff } x \cong_{\mathcal{F}} (x \wedge y) \text{ iff } y \cong_{\mathcal{F}} (x \vee y)$$

but the analogue for $\models_{\mathcal{F}}$ is not generally true. For instance, we have just seen that $(p \leftrightarrow q) \models_{\mathcal{F}} r$ but, of course, r and $(p \leftrightarrow q) \vee r$ are not $\mathcal{F}$-equivalent (the disjunction may be undetermined—or overdetermined, if we were working with $\mathcal{W}$).

This variety of non-equivalent possibilities gives an indication of the degree of sophistication that can be achieved within our generalized semantic framework, where the possibility of gaps and gluts takes us beyond the flatness of the standard landscape. However, there is no need here to further pursue these comparisons. In what follows my concern will be mainly with characterization results, and in such respect our findings will apply equally well to all four notions defined above. Thus, for definiteness, I shall confine myself to $\mathcal{F}$-entailment as defined in (f) unless specified otherwise.

2. DEVELOPMENTS

We have the three fundamental semantic notions: the notion of a language, the notion of a model, and the relative notion of a valuation. I shall now investigate some basic properties of this apparatus. As I mentioned in the Introduction, the emphasis will be on general properties, but I shall also be concerned with a study of the conditions under which some important standard results may be extended to the present, more general account. This should help provide a better understanding of the overall import of the approach pursued here, as well as offer new insights into why those results hold in the standard case.

2.1. VALIDITY

The main properties of our semantic apparatus can be studied with reference to the auxiliary concept of a canonical frame. I shall first focus on some aspects concerning validity. Then I will turn to the more general concept of entailment, and kindred notions.

2.1.1. PRELIMINARIES

Let $\mathcal{F} = \langle K, R, t, \delta \rangle$ be a canonical frame for a language $\mathcal{L}$. An $\mathcal{F}$-valid expression is one that gets the designated value $\delta_{\mathcal{M}}$ in every element $\mathcal{M} \in K$. And since K represents the class of all models deemed "admissible", an expression with that property represents a sort of conceptual invariant—a linguistic item whose interpretation is fixed throughout the conceptual space determined by $\mathcal{F}$. Of course, there is no guarantee that there exist $\mathcal{F}$-valid expressions at all. But if $\mathcal{F}$ is not arbitrarily construed, if it reflects some independently motivated notion of admissibility (some notion of "logical possibility", as we may also say), then the existence of $\mathcal{F}$-valid expressions is a desirable property. In that case we may say that $\mathcal{F}$ is not *vacuous*, or that it is *full*.

Note that, in principle, the notion of $\mathcal{F}$-validity applies regardless of the relevant type t. For example, suppose $t=\langle 0,\langle 0,0\rangle\rangle$. If $\mathcal{L}$ is a sentential language with just one binary connective $\downarrow$, K is the class of all 2-valued, stratified, extensionally adequate models for $\mathcal{L}$, $R=K\times K$, and δ is that constant function such that, for all $\mathcal{M}\in K$ and all $x,y\in\{0,1\}$:

$$\delta_{\mathcal{M}}(x)(y)=1-(x\cup y),$$

then $\downarrow$ is $\mathcal{F}$-valid: its semantic behavior is set by $\mathcal{F}$ to express the truth-conditions of the joint-denial connective 'neither ... nor'. Hence $\downarrow$ becomes a sort of conceptual invariant as far as $\mathcal{F}$ goes, and $\mathcal{F}$ qualifies as a full frame.

In practice, however, the notion of $\mathcal{F}$-validity acquires its traditional interest and significance when $t=0$ and each $\delta_{\mathcal{M}}$ is a designated truth-value—that is, when the relevant semantic category is the category of sentences. In that case, the choice of a particular canonical frame $\mathcal{F}$ may be said to determine a *logic* for the given language $\mathcal{L}$, and the set of all $\mathcal{F}$-valid expressions may be regarded as comprising precisely the *logically valid* sentences of $\mathcal{L}$—those sentences that get the designated truth-value on every admissible way of interpreting them.

It is with this more specific notion of validity that I shall be interested below. First I shall spell out a general fact which—in the light of these remarks—should provide a good characterization of the overall theory of invariance determined by our semantic apparatus. The consequences of this general result for logical validity will then be illustrated with respect to some concrete, familiar cases.

2.1.2. FACT

Let $\mathcal{L}$ be any language and R an ideal admissibility relation on $MOD_{\mathcal{L}}$. If $\mathcal{F}=\langle K,R,t,\delta\rangle$ is a canonical frame for $\mathcal{L}$ and $\mathcal{F}'=\langle K',R',t,\delta'\rangle$, where $K'=K\cap SHRP_{\mathcal{L}}$, $R'=R\upharpoonright K'$, and $\delta'=\delta\upharpoonright K'$, then:

(a) $\mathcal{F}$ is full iff $\mathcal{F}'$ is full; indeed,
(b) for all $x\in\mathbf{E}_t$, x is $\mathcal{F}$-valid iff x is $\mathcal{F}'$-valid.

Proof. (a) is an immediate consequence of (b), so let us prove the latter. To this end, fix $x \in \mathbf{E}_t$ and note that for every $\mathcal{M}' \in K \cap SHRP_{\mathcal{L}}(x)$ we have $\mathcal{V}'_R(x) = \mathcal{V}'_{R'}(x) = f(x)$, where f is the unique x-homomorphism from $\mathcal{L}$ to $\mathcal{M}'$. Now suppose that x is $\mathcal{F}$-valid. Then $\delta_{\mathcal{M}} \in \mathcal{V}_{R,c}(x)$ for all $\mathcal{M} \in K$ and all $c \in C_{\mathcal{M}}$. But $K' \subseteq K$ and $SHRP_{\mathcal{L}} \subseteq SHRP_{\mathcal{L}}(x)$. Thus, in particular, $\delta_{\mathcal{M}'} = \delta'_{\mathcal{M}'} \in \mathcal{V}'_{R,c}(x) = \mathcal{V}'_{R',c}(x)$ for all $\mathcal{M}' \in K'$ and all $c \in C_{\mathcal{M}'}$, i.e., x is $\mathcal{F}'$-valid. Conversely, suppose x is not $\mathcal{F}$-valid. Then there must exist $\mathcal{M} \in K$ and $c \in C_{\mathcal{M}}$ so that $\delta_{\mathcal{M}} \notin \mathcal{V}_{R,c}(x)$, i.e., so that $\delta_{\mathcal{M}} \notin \mathcal{V}''_{R,c}(x)$ for every (x,R)-constriction $\mathcal{M}''$ of $\mathcal{M}$. Pick such an $\mathcal{M}''$ (the existence of which follows from the hypothesis that R is ideal), fix a well-ordering W on $EXP_{\mathcal{L}}$ so that x is W-minimal, and define a family of K-models $A = \{\mathcal{M}_y : y \in EXP_{\mathcal{L}}\}$ such that, for all $y \in EXP_{\mathcal{L}}$: (i) if $x=y$, then $\mathcal{M}_y = \mathcal{M}''$, and (ii) if xWy, then $\mathcal{M}_y$ is a restriction of $\mathcal{M}''$ which is $\sqsubseteq$-maximal in $\bigcap\{CONS_{\mathcal{L}}(z) \cap R[\mathcal{M}_z] : z=y \text{ or } zWy\}$. Intuitively, each $\mathcal{M}_y$ constitutes the yth term of a chain of progressive constrictions of $\mathcal{M}''$. Since every chain $(Z, \sqsupseteq)$ with $Z \subseteq A$ has an $\sqsupseteq$-upper bound in A (just take the model $\sqcap Z$), A will contain an $\sqsupseteq$-maximal element (by Zorn's Lemma). By construction, this element will be a consistent model in K—indeed a constriction $\mathcal{M}^{**} \preccurlyeq \mathcal{M}''$. And since we must have $\mathcal{V}''_{R,c}(x) = \mathcal{V}^{**}_{R,c}(x)$, and consequently $\delta_{\mathcal{M}} \notin \mathcal{V}^{**}_{R,c}(x)$, we can now repeat a perfectly similar argument to find an (x,R)-completion $\mathcal{M}^*$ of $\mathcal{M}^{**}$ and a completion $\mathcal{M}' \in K$ of $\mathcal{M}^*$ such that $\delta_{\mathcal{M}} \notin \mathcal{V}^*_{R,c}(x) = \mathcal{V}'_{R,c}(x)$. But such an $\mathcal{M}'$ will be consistent and complete, hence sharp, hence an element of K'. And in that case we know that $\mathcal{V}'_{R,c}(x) = \mathcal{V}'_{R',c}(x)$. Moreover, each element in the sequence of models considered in the construction of $\mathcal{M}'$ is R-related to its predecessor, and therefore shares with it the same designated value (by definition of canonical frame). This implies that $\delta_{\mathcal{M}'} = \delta_{\mathcal{M}}$. Since $\delta'_{\mathcal{M}'} = \delta_{\mathcal{M}'}$ by assumption, we can therefore conclude that $\delta'_{\mathcal{M}'} \notin \mathcal{V}'_{R',c}(x)$, which means that x is not $\mathcal{F}'$-valid.

2.1.3. REMARKS

In one direction, from left to right, the two biconditionals in 2.1.2 reflect the fact mentioned earlier that, as long as R satisfies certain minimal, intuitive conditions such as reflexivity, the corresponding R-valuations are monotonic. More significant, of course, are the implications in the other direction.

By (a), we see that validity in a canonical frame does not necessarily suffer from the presence of gaps or gluts. A language may contain valid expressions (for instance, sentences which receive a designated truth value in every context of every model) even if some of its admissible models are not sharp. This is because valuations are not fully compositional, and therefore a gap or glut in the pattern of reference of an expression need not result in a gap or glut in the valuation of the expression as a whole: the expression may receive a (designated) semantic value even if its constituents do not. In other words, gaps and gluts (inconsistencies and incompletenesses, underdetermination and overdetermination) are local phenomena that need not "metastasize"[49] throughout a valuation. If, therefore, one were concerned about the idea of allowing for non-sharp models insofar as that might bring logical disaster in its wake, we have here a first, positive result: it is generally not the case that "anything goes" when valuations infringe the standard assumptions of completeness and consistency.

This result is strengthened in (b). For whenever R is an ideal admissibility relation—in the precise sense of Section 1.3.3—the infringement of those assumptions turns out to have no visible effect on the question of *what* counts as a valid expression. Not only is some notion of validity secured, but validity in a canonical frame $\mathcal{F}$ is generally *equivalent* to validity in the sharp fragment of $\mathcal{F}$.[50]

This is still very abstract, and we need consider some concrete examples before drawing any morals. Generally speaking, however, it is important to register the link between these results and the idea of an expression's semantic values being determined by its "explicit" as well as by its "implicit" interpretation, as we put it above. There is but one basic principle expressing this in our semantic theory, namely, that every valuation is the function of a number of sharper valuations. And when this function mirrors a corresponding link between models and sharp models—which is what the present condition on R makes explicit—the upshot is obvious: what holds in every sharp model must hold in every model.

---

[49] The metaphor is from Wittgenstein [1956], Part V, § 8.

[50] This fact extends and generalizes a crucial feature of all supervaluational semantics (compare notes 33 and 34 above).

From this point of view, it is worth mentioning that the above result is liable to several interesting generalizations. In particular, the requirement that R be an ideal admissibility relation may be weakened, as there is in fact a wide spectrum of (more or less intuitively reasonable) relations $R \subseteq K \times K$ for which both (a) and (b) hold. At the one extreme, we have the family of those smallest relations R that associate each model in K with exactly one (x,R)-constriction and exactly one (x,R)-completion for each $x \in EXP_{\mathcal{L}}$. (This would correspond to the view that gluts and gaps are a result of semantic "laziness", as it were, though presumably such a position would make intuitive sense only subject to the condition that gluts and gaps be eliminated in a uniform way for expressions that share common constituents.) At the other extreme, we may take R to be the largest ideal relation on K associating each model with all of its $\sqsubseteq$-maximal x-consistent restrictions and all of its $\sqsubseteq$-minimal x-complete extensions, for each $x \in EXP_{\mathcal{L}}$. No matter which of these indefinitely many relations we take, the fundamental equivalence between $\mathcal{F}$-validity and $\mathcal{F}'$-validity would continue to hold—as one can easily verify with minimal modifications to the initial proof.[51]

Similarly, it is worth remarking that Fact 2.1.2 does not only hold when $\mathcal{F}'$ satisfies the stated conditions. More generally, it can be shown that the equivalences in (a) and (b) hold true even if K' is allowed to be a proper superset of the sharp fragment of K, i.e., they hold whenever $K \cap SHRP_{\mathcal{L}} \subseteq K' \subseteq K$. Thus, although there is a broad spectrum of intermediate positions as to the degree of sharpness to be imposed on a model, and a correspondingly wide spectrum of canonical frames, the resulting logics will not differ as far as validity is concerned. (We shall see that this does not extend to the notion of valid entailment: certain relations of entailment that hold in the sharp fragments do not carry over unrestrictedly to the general case.)

Further interesting generalizations of Fact 2.1.2 can also be obtained by considering different ways of weakening the condition on R. For example, suppose that the identity

$$\mathcal{M} = \bigsqcup\{\bigsqcap\{\mathcal{M}' : \mathcal{M}' \succcurlyeq^{x}_{R} \mathcal{M}''\} : \mathcal{M}'' \preccurlyeq^{x}_{R} \mathcal{M}\}$$

---

[51] This fact can be viewed as a generalization of a point made by van Fraassen [1966b], § 4 (relative to a broad spectrum of supervaluational policies).

only holds with respect to some $x \in EXP_{\mathcal{L}}$. If R is such a relation (call it an *x-ideal* admissibility relation), Fact 2.1.2 does not necessarily apply. Yet one can easily verify that x comes out $\mathcal{F}$-valid iff it is $\mathcal{F}'$-valid, where $K' = K \cap SHRP_{\mathcal{L}}(x)$ (or, more generally, $K \cap SHRP_{\mathcal{L}}(x) \subseteq K' \subseteq K$). Thus, our earlier remark that the evaluation of an expression x on a model $\mathcal{M}$ does not call for a full sharpening of $\mathcal{M}$ (one must consider models that are complete and consistent relative to x) can generally be extended to the notion of validity: *if* the interpretation of x is always sharpenable (in some way or other), then for the purpose of checking the validity of x one can simply rely on the x-sharp fragment of $\mathcal{F}$.

Incidentally, all of this is quite independent of our particular formulation of Definition 1.3.2. The above results carry over easily even if we modify our framework by considering the duals of a language's valuations. Let us now consider some examples.

2.1.4. COROLLARIES

Let $\mathcal{L}$ be any language and let $\mathcal{F} = \langle K, R, t, \delta \rangle$ be a canonical frame for $\mathcal{L}$, where R is ideal, $t=0$, and $\delta_{\mathcal{M}}=1$ for all $\mathcal{M} \in K$. Then:

(A) If $\mathcal{L}$ is a sentential language with a binary connective $\downarrow$, as in Example 1.1.4(A), and K is the class of all 2-valued, extensionally adequate sentential models for $\mathcal{L}$, as in Example 1.2.4(A), then a sentence $A \in \mathbf{E}_0$ is $\mathcal{F}$-valid iff it is rated valid in classical sentential logic.

Proof. It suffices to note that the logic determined by the frame $\mathcal{F}' = \langle K', R', t, \delta' \rangle$, where $K' = K \cap SHRP_{\mathcal{L}}$, $R' = R \restriction K'$, and $\delta' = \delta \restriction K'$, is precisely classical sentential logic: given any $\mathcal{M}' \in K'$, the corresponding R'-valuation maps each $\mathcal{L}$-sentence to a function $C_{\mathcal{M}'} \rightarrow \{0,1\}$ with the sole proviso that $\mathcal{V}'(A \downarrow B)(c) = 1 - (\mathcal{V}'(A)(c) \cup \mathcal{V}'(B)(c))$ (see Example 1.3.4(A)). Hence the result follows directly from Fact 2.1.2.

Remarks. Since a valuation on a 2-valued, extensionally adequate sentential model $\mathcal{M}$ always satisfies the biconditional: $z \in \mathcal{V}_c(\neg A)$ if and only if $1 - z \in \mathcal{V}_c(A)$, a perfectly similar argument will also show that the set of all sentences taking the value 0 in every context of every $\mathcal{F}$-model is exactly the set of all classical contradictions. Indeed, we get a full picture of the logic of $\mathcal{F}$ by observing that classical tautologies and contradictions are always sure to be single-valued, that is, also on incomplete and inconsistent valuations. (This is a by-product of the general proof given earlier: whenever $z \in \mathcal{V}_c(A)$ for some $\mathcal{M} \in K$ and some $c \in C_{\mathcal{M}}$, we must have $z \in \mathcal{V}'_c(A)$ for some $\mathcal{M}' \in K \cap SHRP_{\mathcal{L}}$.) Accordingly, tautologies are always true and never false, and contradictions are always false and never true.

One consequence of this, of course, is that even sentences such as the following retain their traditional logical status:

(1) $A \vee \neg A$

(2) $A \wedge \neg A$.

This relates to our previous observation (in example 1.3.4(A)) that sentences of this form are not directly related to the semantic principles of completeness and consistency—namely, to the statements that a sentence A must always be either true or false, and cannot be both true and false. If we confine ourselves to sharp models, these principles are indeed mirrored in the semantic status of (1) and (2). But in general this is not true. As we have seen, from the fact that $1 \in \mathcal{V}_c(A \vee \neg A)$ we cannot conclude that either $1 \in \mathcal{V}_c(A)$ or $1 \in \mathcal{V}_c(\neg A)$, for $\mathcal{V}_c(A)$ might be undefined; and from the fact that $0 \in \mathcal{V}_c(A \wedge \neg A)$ we cannot exclude that $1 \in \mathcal{V}_c(A)$ and $1 \in \mathcal{V}_c(\neg A)$, for $\mathcal{V}_c(A)$ might be overdefined.[52]

There is of course room for discussion here, and I do not intend to claim that this is *the* correct generalization of classical 2-valued sentential logic: alternate intuitions may be modeled by choosing different (canonical) frames for $\mathcal{L}$. Arguably, however, this position seems fair if our concern is with epistemic gaps and gluts, rather than full-blooded, on-

---

[52] The distinction between the logical law expressed by (1) ("excluded middle") and the semantic requisite of completeness ("bivalence") was first emphasized by Mehlberg [1956], §29, and (independently) by McCall [1966] and especially van Fraassen [1966a], §8. Horwich [1990] and Williamson [1992, 1994] have argued against the distinction, but their reasoning seems to me to be adequately rebutted in Simons [1992] and Wright [1995]; see also Griffin [1978] vs. Machina [1976], and Day [1992] vs. Sayward [1989]. A different line of objection may be found in Kripke [1975a], Sanford [1976], and Tye [1989, 1990] among others, who reject the distinction to save compositionality for all connectives. The case for (2) can be traced back to Jaśkowski [1948] and Przełęcki [1964], but see also Rozeboom [1962]; more recent sources are Belnap [1977], Rescher & Brandom [1980], Lewis [1982], and Urchs [1994]. I discuss it at some length in Varzi [1997].

tological incompleteness and inconsistency. Just as the disjunction or conjunction of two opposite events, each having probability less than 1 and greater than 0, can have a definite probability (1 or 0), so the disjunction or conjunction of a sentence and its negation may be regarded as having a definite truth value even when the values of its constituents are not defined, or not uniquely defined.[53] An unsharp model need not mirror an unsharp world. (Think again of the Holmes stories: there are lacunae and discrepancies, but their model of the world is no logical chaos.) If gaps and gluts were truly ontological—if at least some of them resulted from there being defective or incongruous objects or states of affairs, as some have suggested[54]—then one could not quarantine them this way. But that would amount to saying that some models are not sharpenable at all, hence that R is not an ideal admissibility relation. And in such circumstances Fact 2.1.2 simply does not apply.

Note that all of this is subject to the generalizations mentioned in the previous section. For instance, we could take K to include only consistent models, or only complete models, or any mixture of such models: as long as all sharp 2-valued, extensionally adequate sentential models are included (up to isomorphism), the equivalence between $\mathcal{F}$-validity and validity in classical sentential logic is secured.

It is also worth observing what happens if we take K to consist of n-valued extensionally adequate sentential models for $\mathcal{L}$, with $n>2$. In that case, it is easy to verify that the natural choice of δ, with $\delta_{\mathcal{M}}=n-1$ for all $\mathcal{M}\in K$, would determine an empty set of $\mathcal{F}$-valid sentences. However this is not due to a sudden overflowing of inconsistency or incompleteness: it is a fact that even if we restrict ourselves to sharp n-valued mod-

---

[53] Arguments of this sort have been offered by most authors cited in note 52 (e.g. Fine [1975], p. 270, or Przełęcki [1982]). The case for $A\wedge\neg A$ seems perfectly dual (as in the "preface paradox" of Makinson [1964]); see also Kyburg [1970, 1997]. Epistemic semantics exploiting these intuitions can be found in Levesque [1984, 1990], Konolige [1985], Fagin & Halpern [1987] (see Vardi [1986] for connections).

[54] An ontology with incomplete and inconsistent entities has been defended *inter alia* by Parsons [1980], Routley [1980], and other authors inspired by the theory of Meinong [1904]. Incomplete or otherwise defective objects have also been considered by authors concerned with vagueness: see e.g. van Inwagen [1988], Tye [1990], Zemach [1991], and Parsons & Woodruff [1995]. On the contrast between vague ontology and supervaluational semantics, see also Heller [1990].

els, the corresponding notion of validity will remain empty, since every sentence could be assigned a value distinct from $n-1$. Indeed, consider the following alternative notion of $\mathcal{F}$-validity:

> A is *negatively $\mathcal{F}$-valid* iff, for all $\mathcal{M} \in K$ and all $c \in C_{\mathcal{M}}$, $0 \notin \mathcal{V}_c(A)$.

This amounts to negative $\mathcal{F}$-validity (i.e., negative $\mathcal{F}$-entailment by $\varnothing$) with 0 as the anti-designated value. Then one can show that the negatively $\mathcal{F}$-valid sentences of $\mathcal{L}$ are exactly the classic tautologies—those sentences that are never false. More generally, consider the definition:

> A is *weakly $\mathcal{F}$-valid* iff, for all $\mathcal{M} \in K$ and all $c \in C_{\mathcal{M}}$, $i \in \mathcal{V}_c(A)$ for some $i \geq \frac{n-1}{2}$.

(This is a stronger notion than negative $\mathcal{F}$-validity, though weaker than $\mathcal{F}$-validity.) Then one can show that a sentence of $\mathcal{L}$ is weakly $\mathcal{F}$-valid just in case it is weakly $\mathcal{F}'$-valid, where $K' = K \cap SHRP_{\mathcal{L}}$. And it is a known fact that no matter which $n>2$ we pick, an $\mathcal{L}$-sentence is weakly $\mathcal{F}'$-valid just in case it is rated valid in classical sentential logic.[55]

From this point of view, the present framework provides a generalization of the familiar connection between classical sentential logic and its weakly equivalent many-valued extensions.[56] Indeed Fact 2.1.2 allows us to go further, as reference to classical logic here is purely illustrative. If instead of extensionally adequate models we took models that give the binary connective $\downarrow$ a different interpretation, then the (weakly) $\mathcal{F}$-valid sentences would not coincide with the classical tautologies. They would, however, coincide with those sentences that are (weakly) satisfied throughout the relevant class of sharp models, whatever that might be. The general point is not that $\mathcal{F}$-validity reduces to classical validity, but that it reduces to sharp validity. This will be clearer in the next example.

(B) If $\mathcal{L}$ is an elementary language with a set $V_{\mathcal{L}}$ of name variables and a binary quantifier $\langle\downarrow, v\rangle$ for each $v \in V_{\mathcal{L}}$, as in Example 1.1.4(B), and if K is the class of all 2-valued, non-empty, logically adequate elementary models for $\mathcal{L}$, as in Example 1.2.4(B), then a sentence $A \in \mathbf{E}_0$ is $\mathcal{F}$-valid iff it is rated valid in classical elementary logic.

---

[55] See Rescher [1969].

[56] See Herzberger [1982], § 11, for some hints in this direction.

Proof. Again, it suffices to note that the elementary logic determined by the frame $\mathcal{F}'=\langle K',R',t,\delta'\rangle$, where $K'=K\cap SHRP_{\mathcal{L}}$, $R'=R\upharpoonright K'$, and $\delta'=\delta\upharpoonright K'$, is classical (see Example 1.3.4(B)). The result then follows immediately from Fact 2.1.2.

Remarks. The above remarks on sentential logic can all be extended —*mutatis mutandis*—to the present case. There are, however, some other facts worth of notice.

To begin with, observe that not only such tautologous or contradictory sentences as (3) and (4) retain their classical logical status in $\mathcal{F}$, but even sentences like (5) and (6) do so:[57]

(3) $Px\vee\neg Px$
(4) $Px\wedge\neg Px$
(5) $Px\vee\neg\bigwedge vPv$
(6) $Px\wedge\neg\bigvee vPv$

Sentence (5) is never false and always true (i.e., it gets the unique value 1 in every context of every model in K), while (6) is never true and always false (it always gets the unique value 0). Again, this means that such sentences do not express completeness and inconsistency, contrary to what it might seem. Yet in this case two complementary factors are intertwined in a delicate way. On the one hand, incompleteness or inconsistency may be due to the presence of an undefined or overdefined predicate P (for instance, a predicate whose denotation is an item $x\in\mathbf{I}_{\langle 1,0\rangle}$ such that $\mathbf{h}(x,y)[c]=\varnothing$ or $\mathbf{h}(x,y)[c]=\{0,1\}$ for some $y\in\mathbf{I}_1$ and some $c\in C_M$).[58] In this case, the logical status of (3)–(6) stems directly from the potential or implicit meaning of P, matching our earlier example: just as in a 2-valued sentential model a sentence can only be granted one of two possible truth-values in each context, in a 2-valued elementary model a predicate can only be sharpened in one of two possible ways with respect to each argument. On the other hand, the source of incom-

---

[57] Here and below I shall rely on the notational conventions of Chapter 1, but I shall omit unnecessary parentheses.

[58] Note that a lack of denotations for 'P' in a context c may be identified with a denotation assignment that is totally undefined in c—i.e., an assignment $x\in\mathbf{I}_{\langle 1,0\rangle}$ such that $\mathbf{h}(x,y)[c]=\varnothing$ for all $y\in\mathbf{I}_1$. Likewise for an abundance of denotations. See above, note 18.

pleteness or inconsistency may also lie in a lack or an abundance of denotations for a name x. And in this case, such sentences as (5) or (6) appear more problematic, due to the intuitive interpretation of the quantifiers. For instance, the truth of (5) is seemingly in contrast with the possibility that x has no denotation in a model $\mathcal{M} \in K$, for such a sentence suggests that x is a legitimate substitute of the bound variable v. Even worse, suppose that $FUN_{\mathcal{L}}$ contains a binary predicate $\mathbf{s}(\beta,t) = \equiv$ that is interpreted in a model $\mathcal{M}$ as the identity relation (as in example 1.3.4(B)). Then also a sentence such as

(7) $\bigvee v(x \equiv v)$

would come out true in $\mathcal{M}$, even if x has no denotation. And this sentence *says* that x is a legitimate substitute of a bound variable.

Once again, the reason for all this lies in the potential meaning of x as set by the canonical frame $\mathcal{F}$. Recall that a model's extensions are always bound to have the same context set as the model itself (by definition). Since $\mathcal{F}$ is based on an ideal admissibility relation on K, and since every element $\mathcal{M} \in K$ must satisfy the condition $C_{\mathcal{M}} = \mathbf{D}_1{}^{V_{\mathcal{L}}}$, the completions of an incomplete model in K are therefore bound to have the same basic domains of interpretation as the model itself—that is, not just the same $\mathbf{D}_0 = 2$, but also the same $\mathbf{D}_1$. This implies that a lack of denotation for a name x in a context c of a model $\mathcal{M}$ can only be redeemed by an element of $\mathbf{U}_1$. Hence, to say that x is a legitimate substitute of a bound variable amounts in the present setting to saying that x can only be granted one of a fixed number of possible denotations in each context, all of which are already available in the given domain. And this restores the symmetry between (7) and '$A \vee \neg A$' (where A is truth-valueless) mentioned in Example 1.3.4(B): in both cases, we have a compound sentence that is true by virtue of a truth that is preserved for all possible ways of interpreting its undefined constituents.[59]

From this perspective, the situation portrayed above reflects the view that truth-value gaps (gluts) are but one instance of a general sort of semantic incompleteness (inconsistency): denotation gaps or gluts for

---

[59] See again Fine [1975] or Dummett [1975]. Misgivings on this treatment of quantifiers have been manifested by Kripke [1975a] and Sanford [1976] and, more recently, by Rolf [1984], Burns [1991], and Tappenden [1993] among others.

names are incidents of that same sort, and must therefore be accounted for in the same way. This does not mean, however, that differing views are incompatible with our basic framework. For example, to obtain a logic that validates all classical tautologies but not such quantified sentences as (5) or (7) we may rely on a different (non-canonical) frame. Suppose we extend our relation $R \subseteq K \times K$ to a relation $R' \subseteq K \times K'$, where K' consists of those models $\mathcal{M}'$ that are just like the members of K except that, in general,

$$C_{\mathcal{M}'} = X^{V_{\mathcal{L}}} \text{ for some } X \subseteq \mathbf{D}_1'.$$

In other words, suppose we allow a model $\mathcal{M}$ to have admissible completions $\mathcal{M}'$ which, though provided with the same domain of quantification as $\mathcal{M}$ itself (in view of the identities $C_{\mathcal{M}'} = C_{\mathcal{M}} = \mathbf{D}_1{}^{V_{\mathcal{L}}}$) may interpret name constants by elements *foreign* to that domain (i.e., by elements of $\mathbf{D}_1' - \mathbf{D}_1$). The interpretation of a quantifier $\mathbf{s}(\beta,t) = \langle \downarrow, v \rangle$ on such a completion becomes:

$$\mathbf{h}'(\mathbf{h}'(\mathbf{d}'(\beta,t),x),y)(c) = 1 \text{ iff } x(c(v/u)) \cup y(c(v/u)) = 0 \text{ for all } u \in \mathbf{D}_1.$$

Then, if R' is still ideal and $\mathcal{F} = \langle K, R', t, \delta \rangle$, the resulting notion of validity is essentially that of a logic in which name constants do not carry existential presuppositions: (5) and (7) will hold unrestrictedly if, but only if, x is a variable.[60] (Intuitively, these conditions on the admissibility relation reflect the idea that a completion need not interpret non-denoting names by means of "existents". A somewhat stronger attitude would be that a completion *must* not interpret non-denoting names by existents. To account for this stronger view, one may add the requirement that $\mathcal{M}R'\mathcal{M}'$ only if the following condition is satisfied for all $\langle \beta,t \rangle \in \mathcal{D}s$:

$$\text{if } \mathbf{d}(\beta,1)[c] = \varnothing, \text{ then } \mathbf{d}'(\beta,1)[c] \cap \mathbf{D}_1 = \varnothing.$$

In both cases, a gap in the denotation of x may imply a gap in the value of (5). However, if $\mathcal{M}$ interprets $\equiv$ as the identity relation, then (7)

---

60 So-called free logics (see note 16) are usually formulated in languages with a distinguished identity or existence predicate. See Lambert [1963], Leblanc & Meyer [1970], Fine [1983], and Bencivenga [1986] for "pure" formulations that do not require such constraints.

would come out gappy on the first account, and false on the second.[61] For future reference, I shall distinguish these two accounts by calling R' a *free* or *strongly free* variant of R, respectively.)

The difference between these two alternatives—between a logic that validates (5) or (7) and one that does not—is an important one. It reflects different senses in which we may take a name to lack a denotation (in a context). On the first view, a name may lack a denotation only in the sense in which a sentence symbol typically lacks a denotation: the model does not specify *which* denotation it has. The name may be undefined (the stories don't say what date corresponds to 'Watson's birth date') or vaguely defined (each one of a large variety of slightly distinct and precisely determinate aggregates of molecules has an equal claim to being the denotation of 'Mount Everest').[62] In such cases, the truth of (5) and (7) is warranted. So if these are the only cases, (5) and (7) are to come out logically true. This is what our Corollary says. On the second view, a name may also lack a denotation insofar as it aims to pick out an entity that is not—or that may not be—in the domain of quantification ('Pegasus'). When it comes to sharpening the model, it would therefore be inappropriate to consider only sharpenings based on the available domain: other ways of interpreting the name must also be considered. And this amounts to considering a free variant of the given accessibility relation R (or a strongly free variant, if one insists that these other ways are the only acceptable ways of interpreting the name).[63]

Incidentally, note that we would *have* to go free if we allowed K to include empty models, i.e., models $\mathcal{M}$ with $\mathbf{D}_1 = \varnothing$. Otherwise $\mathcal{F}$ could be vacuous. More precisely, the only $\mathcal{F}$-valid sentences (if any) would belong to the sentential reduction of $\mathcal{L}$. This is because every extension $\mathcal{M}' \in K$ of a model $\mathcal{M} \in K$ must satisfy a twofold requirement—namely: (i) the context set of $\mathcal{M}'$ must coincide with that of $\mathcal{M}$ (by definition

[61] The second, "stronger" attitude is apparently more popular among free logicians, who naturally identify denotationlessness with non-existence. See van Fraassen [1966a], § 5, for a first explicit statement of this view in semantic terms. By contrast, the "weak" attitude may be traced back to Skyrms [1968].

[62] This was the gist of Mehlberg's treatment of vagueness in [1956], §29. See also Quine [1985], pp. 167–68.

[63] For more on the philosophical underpinnings of free logic, in the sense relevant here, see Bencivenga [1990].

of $\sqsupseteq$), hence it must coincide with the set $\mathbf{D}_1{}^{V_{\mathcal{L}}}=\{\varnothing\}$; and (ii) the domain $\mathbf{D}_1'$ must coincide with the range of those functions (by definition of $\mathcal{K}$), hence it must coincide with the set $\mathbf{D}_1=\varnothing$. Thus, an empty model $\mathcal{M}$ would have an empty class of (A,R)-completions for every $A\in\mathbf{E}_1$ whose constituents include names (constant or variable), yielding a corresponding gap $\mathcal{V}_c(A)=\varnothing$. By contrast, suppose we go for a free admissibility relation. More specifically, suppose we now extend our relation $R\subseteq\mathcal{K}\times\mathcal{K}$ to a relation $R'\subseteq\mathcal{K}\times\mathcal{K}'$, where $\mathcal{K}'$ comprises those models $\mathcal{M}'$ that are just like the members of $\mathcal{K}$ except that (i) $C_{\mathcal{M}'}$ consists of assignments taking values in an arbitrary subset $X\subseteq\mathbf{D}_1'$ (as above), and (ii) when this set X is empty, i.e., when $C_{\mathcal{M}'}=\varnothing^{V_{\mathcal{L}}}=\{\varnothing\}$, the denotation of a name variable $\mathbf{s}(\beta,t)=v$ is emended accordingly, setting

$$\mathbf{d}'(\beta,t)(\varnothing)\in\mathbf{D}_1'.$$

Then nothing prevents an empty model $\mathcal{M}$ from having non-empty (A,R')-completions even when A includes names, and R' may be A-ideal. The Corollary would fail to apply, of course, for the resulting frame would not be canonical. But the corresponding notion of validity would not be trivial. In particular, (5) and (7) may now fail even when x is a variable, and a classically valid sentence such as

(8) $\bigvee v(v\equiv v)$,

would fail too. Nonetheless, the universal closures of (5) and (7),

(5') $\bigwedge w(Pw\vee\neg\bigwedge vPv)$
(7') $\bigwedge w\bigvee v(w\equiv v)$,

as well as sentences such as

(9) $v\equiv v$
(10) $\bigwedge v(v\equiv v)$

would be valid. These sentences—we may say—would express statements whose truth does not depend on there being anything at all in the domain of quantification.[64] (For instance: whereas (7) says that x is

[64] The outcome would be a logic that is both free and "inclusive", in sense of note 15—what is sometimes known as a "universally free logic" (Meyer & Lambert [1968]).

identical with something in the domain, i.e., that x exists, (7') says that everything in the domain is identical with something in the domain, i.e., that everything that exists exists—and that is naturally regarded as true even in the empty context of an empty model.)

Various other aspects of the Corollary tie in with this issue. As a last illustration, suppose $\mathcal{L}$ includes, besides perhaps an identity predicate, a stock of subnectives to be interpreted as definite descriptors (corresponding to the English article 'the', in the singular). Within the present framework these can be treated in a way similar to other variable binders: we may assume $SUB_{\mathcal{L},1}$ to include a structured symbol $\mathbf{s}(\beta,t)=\langle\iota,v\rangle$ for each $v\in V_{\mathcal{L}}$; and we may say that a model $\mathcal{M}$ interprets $\langle\iota,v\rangle$ as the *definite descriptor* binding variable v just in case, for all $x\in\mathbf{I}_0$ and all $c\in\mathbf{D}_1{}^{V_{\mathcal{L}}}$,

$$\mathbf{h}(\mathbf{d}(\beta,t),x)[c]=\{u\in\mathbf{D}_1\colon x(c(v/u))=1\}.$$

(In other words, $\mathcal{M}$ must interpret $\langle\iota,v\rangle$ as that operator which, when applied to a propositional function x, picks out an object u on an assignment c if and only if x yields the value "true" on the assignment $c(v/u)$.) Then it is easy to show that the above Corollary does not hold if we restrict K to elementary models of this sort. The reason, quite simply, is that there is no classical elementary logic for such a language, as the completeness condition for $\mathbf{h}(\mathbf{d}(\beta,t),x)[c]$ can always be falsified—for instance, by taking x so that $x(c(v/u))=0$ for every $u\in\mathbf{D}_1$. It is precisely for this reason that definite descriptors are usually banned from the vocabulary of standard elementary languages.[65] But the point is —more generally—that if we restrict K by interpreting descriptors as indicated, then no admissible model can be sharpened with respect to every sentence of the language. For consider a "contradictory" description such as

(11) $\iota v(Qv\wedge\neg Qv)$

(in the obvious notation). We have seen that a conjunction of the form $Qv\wedge\neg Qv$ can never be true in a K-model. This means that every sen-

---

[65] The standard account is of course that of Russell [1905]; for a recent defense, see Neale [1990]. For an overview of the problems, see Ostertag [1998].

tence A whose constituents include (11) will have no A-sharpenings in $\mathcal{K}$. This in turn implies that the basic identity

$$\mathcal{M} = \bigsqcup\{\bigsqcap\{\mathcal{M}': \mathcal{M}' \succcurlyeq^A_R \mathcal{M}''\}: \mathcal{M}'' \preccurlyeq^A_R \mathcal{M}\}$$

will fail for every $\mathcal{M} \in \mathcal{K}$. (We may say that in such cases A turns out to be *essentially* undefined[66] in every $\mathcal{M} \in \mathcal{K}$: its lack of values is not due to a disagreement of the relevant completions of (constrictions of) $\mathcal{M}$, but to the *nonexistence* of such completions.) Hence Fact 2.1.2 does not apply, and Corollary 2.1.4(B) does not follow.

Nor are contradictory descriptions the only case in point. It is clear that the same applies whenever A involves descriptions that are mutually incompatible, such as

(12) ιvQv
(13) $\iota v(Qv \wedge Sv)$
(14) $\iota v(Qv \wedge \neg Sv)$

even if each one of them is individually liable of being assigned a sharp interpretation under suitable circumstances. Conclusion: in all these cases, even sentences of the form (3)–(6) may lose their logical status. Indeed, since every sentence has a denumerable variety of essentially undefined substitution instances, involving contradictory or mutually incompatible definite descriptions, it follows that in a language with descriptors a substitution instance of a valid elementary sentence need not be valid.[67] And, conversely, since every description can be used to ob-

---

[66] This terminology is borrowed from Bencivenga [1978, 1980c], whose semantics for definite and indefinite descriptions inspired the present remarks.

[67] A rigorous characterization is as follows. Let $\mathcal{L}_1$ be the initial elementary language, without descriptions, and define $\mathcal{L}_2 \succeq \mathcal{L}_1$ by adding a descriptor $\langle\iota,v\rangle$ for each name variable v of $\mathcal{L}_1$. Next, extend the notion of a free variable (note 10) to cover the case of descriptors in the obvious way. For all expressions x,a,b of $\mathcal{L}_2$, define the expression $x^a/_b$ resulting from substituting a for b in x, as in 1.3.4(C): (i) if $x=b$, then $x^a/_b=a$; (ii) if $x=\mathbf{g}(y,z)$ and $y\neq\langle H,v\rangle$ for each $v\in V_{\mathcal{L}_2}$, where $H\in\{\iota,\downarrow\}$, then $x^a/_b=\mathbf{g}(y^a/_b,z^a/_b)$; (iii) if $x=\mathbf{g}(y,z)$ and $y=\langle H,v\rangle$ for some $v\in V_{\mathcal{L}_2}$, where $H\in\{\iota,\downarrow\}$ and $v\neq b$, then $x^a/_b=\mathbf{g}(\langle H,v\rangle,z^a/_b)$ if v does not occur free in a or b does not occur free in z, and $x^a/_b=\mathbf{g}(\langle H,w\rangle,z^w/_v{}^a/_b)$ otherwise, where w is the first element of $V_{\mathcal{L}_2}-(\omega_{\mathcal{L}_2}(a)\cup\omega_{\mathcal{L}_2}(z)\cup\{v,b\})$; (iv) $x^a/_b=x$ in all other cases. Given these definitions, let K_i be the class of all 2-valued, logically adequate elementary models for $\mathcal{L}_i$ (i=1, 2), and set $\mathcal{F}_i=\langle K_i,R_i,0,\delta_i\rangle$, with $R_i\subseteq K_i\times K_i$ and $\mathcal{R}\delta_i=\{1\}$. Then the point is that for every $\mathcal{F}_1$-

tain an essentially undefined substitution instance of a valid sentence (just take a sentence of the form

(15) $Pxx_1 \dots x_n \vee \neg Pxx_1 \dots x_n$,

where x is the description in question and $x_1, \dots, x_n$ are contradictory or mutually incompatible descriptions), it also follows that no description conforms to all the logical truths of a description-free elementary language.

The picture looks even worse if we consider that, as things are, this "loss" is not compensated by a discovery of any new patterns of validity, characteristic of a language with descriptions. We have seen that the basic characterization of a model in terms of A-completions and A-constrictions fails when A is a sentence involving contradictory or mutually incompatible descriptions. But inspection shows that the same applies whenever A involves descriptions that are simply *non-denoting* in some context (i.e., on some assignment of values to the name variables)—even a single, perfectly innocent description such as (12). Once again, this depends on the fact that the value assignments making up the context set of an elementary model $\mathcal{M} \in K$ consist of functions ranging on the basic domain $\mathbf{D}_1$. Accordingly, a description which does not have any value on an assignment $c \in C_{\mathcal{M}}$ is sure to remain such no matter how we expand $\mathcal{M}$. For instance, if $\mathcal{V}_{c(v/u)}(Qv)=\{0\}$ for all $u \in \mathbf{D}_1$, then every admissible consistent extension $\mathcal{M}'$ of $\mathcal{M}$ will agree on the same values $\mathcal{V}'_{c(v/u)}(Qv)=\{0\}$ (by monotonicity) and will therefore remain incomplete with respect to the description ιvQv—hence with respect to any sentence whose constituents include ιvQv. Nor would things change if we took R to be a (strongly) free admissibility relation: for although in that case the extensions of $\mathcal{M}$ might allow for new objects which *are* Q, those objects will not be in the range of interpretation of $\mathcal{L}$'s name variables, and ιvQv will still remain denotationless in c.

Of course, there is a sense in which this situation is perfectly reasonable: one could observe that the denotationlessness of a description of the form ιvQv does not quite reflect a "gap" of $\mathcal{M}$ but, rather, the un-

---

valid sentence A of $\mathcal{L}_1$ there exist denumerably many substitution instances of A that are not $\mathcal{F}_2$-valid. Moreover, it is easy to verify that the same applies even if we redefine $\mathcal{F}_1$ and $\mathcal{F}_2$ using (strongly) free variants of R_1 and R_2, respectively.

ambiguous fact that no object in $\mathbf{D}_1$ satisfies the predicate Q (i.e., that Qv is false on every assignment of a value $u \in \mathbf{D}_1$ to the name variable v). There are indeed various factors at issue here. Once again, however, the point to be stressed in the present context is that these results stem from defining validity in a certain way, with respect to a canonical frame satisfying certain conditions. The picture would change if a different frame were chosen instead of $\langle K, R, 0, \delta \rangle$, where R is an ideal admissibility relation on K (or a (strongly) free variant thereof).

To illustrate, suppose that we extend R to a new relation $R' \subseteq K \times K'$, where K' consists of those models $\mathcal{M}'$ that are just like the members of K except for the interpretation of the descriptor subnectives; where $\mathbf{s}(\beta,t)=\langle \iota, v \rangle$, the following condition applies instead:

$$\mathbf{h}'(\mathbf{d}'(\beta,t),x)[c] \supseteq \{u \in \mathbf{D}_1' : x(c(v/u))=1\}.$$

This is weaker than the condition imposed on the members of K. In particular, a model $\mathcal{M} \in K$ may now have an admissible extension $\mathcal{M}' \in K'$ which treats a description x as denoting an element u even if u does not "fit" x—which is enough to guarantee completability where appropriate. Then it is obvious that if R' is sufficiently tolerant, the resulting (non-canonical) frame $\mathcal{F}=\langle K, R', t, \delta \rangle$ will allow for valid sentences with descriptions. For instance, the result of putting a sharpenable description in place of the name x in (3)–(6) will yield an $\mathcal{F}$-valid sentence in each case. Indeed, if we take R' to be an ideal relation, then a sentence $A \in \mathbf{E}_0$ turns out to be valid in this frame if and only if it is a substitution instance of a sentence rated valid in classical elementary logic (without descriptions).[68] And if we took an ideal free variant of R' instead, obtained by extending K' in the obvious way, then the resulting notion of $\mathcal{F}$-validity would coincide with validity in a system of free logic (without descriptions).[69]

---

[68] If we required completions to assign the same, distinguished value to every non-denoting description (for instance, the first element relative to some canonical well-ordering of $\mathbf{D}_1'$), the resulting account would be closely related to the "chosen object" theory of Carnap [1947], which follows in the footsteps of Frege [1893, 1903]. A related account is that of von Kutschera [1975], where the distinguished object is identified with the value of a fixed name constant.

[69] This account would bear some connections with the free semantics of van Fraassen & Lambert [1967].

It is also easy to consider additional conditions on R' which would determine patterns of validity characteristic of sentences containing definite descriptions. For example, suppose we add a requirement to the effect that $\mathcal{M}R'\mathcal{M}'$ only if $\mathcal{M}'$ interprets non-denoting descriptions as elements foreign to $\mathbf{D}_1$, in the spirit of a strongly free variant of R—formally:

$$\text{if } \mathbf{h}(\mathbf{d}(\beta,t),x)[c]=\varnothing, \text{ then } \mathbf{h}'(\mathbf{d}'(\beta,t),x)[c]\cap\mathbf{D}_1=\varnothing.$$

And suppose every model in the range of R' interprets $\equiv$ as the identity relation. Then it is easy to verify that sentences of the form

(16) $\neg\bigvee vA \rightarrow \neg\bigvee w(w\equiv \iota vA)$
(17) $\neg\bigvee w(w\equiv \iota v(Qv\wedge\neg Qv))$

would come out valid in the resulting frame. They would express the (conditional and unconditional) non-existence of a designatum for the relevant descriptions.[70]

2.2. ENTAILMENT

Validity is a special case of entailment, understood as a relation between sets of expressions. I shall now turn to this more general notion and compare more thoroughly its standard behavior with the effects of embracing incompletenesses and inconsistencies as *bona fide* semantic possibilities.

2.2.1. PRELIMINARIES

Let $\mathcal{F}=\langle K,R,t,\delta\rangle$ be a canonical frame for a language $\mathcal{L}$. If a set of expressions X is $\mathcal{F}$-entailed by a set of expressions Y (i.e., if $Y \vDash_{\mathcal{F}} X$), we may say that the *argument* $Y\vdash X$ is *$\mathcal{F}$-valid*, referring to Y and X as the

[70] This account could be further strengthened by requiring $\mathbf{D}_1'-\mathbf{D}_1$ to comprise a single distinguished element u^*, in the spirit of Scott [1963]. (Compare Blamey [1986], § 6.3.) The outcome would be a "free" version of the chosen object theory mentioned in note 68.

premisses and the *conclusions* of the argument, respectively. (This is of course quite general: in fact the term "argument" acquires its customary meaning only in connection with sets of sentences.) Evidently, for every $\mathcal{F}$-valid expression x there is a matching $\mathcal{F}$-valid argument $\varnothing \vdash \{x\}$, and we have seen that whenever R is an ideal admissibility relation, the $\mathcal{F}$-valid expressions coincide exactly with the valid expressions in the sharp fragment of $\mathcal{F}$ (Fact 2.1.2). However, a similar fact does not generally hold for argument validity. For example, if $\mathcal{L}$ is a sentential language and $\mathcal{F}$ and $\mathcal{F}'$ are as in Corollary 2.1.4(A), then such classical arguments as

(18) $X \vdash A, \neg A$
(19) $A, \neg A \vdash X$

are $\mathcal{F}'$-valid, but not necessarily $\mathcal{F}$-valid:[71] (18) fails whenever A can be undetermined, and (19) fails when A can be overdetermined. Or suppose that $\mathcal{L}$ is a typical elementary language with quantifiers and $\mathcal{F}$ and $\mathcal{F}'$ are as in Corollary 2.1.4(B). Then, again, arguments such as

(20) $X \vdash Px, \neg \bigwedge v Pv$
(21) $Px, \neg \bigvee v Pv \vdash X$

are $\mathcal{F}'$-valid, but they may lose their status when it comes to $\mathcal{F}$-validity: the predicate P may be undefined or overdefined (respectively); or its denotation may be undefined or overdefined (respectively) for some arguments.

This asymmetry is of course crucial, particularly when—as in the examples above—t is the type of sentences and δ_M some designated truth-value. For the upshot is that gaps and gluts may not affect the logically valid sentences of a language, but they surely make a difference with respect to the logically valid *arguments*. To be more precise, standard valid arguments still prove adequate when it comes to drawing

---

[71] If $y_1, \ldots, y_m, x_1, \ldots, x_n$ are expressions, the notation '$y_1, \ldots, y_m \vdash x_1, \ldots, x_n$' will abbreviate '$\{y_1, \ldots, y_m\} \vdash \{x_1, \ldots, x_n\}$'. More generally, if $Y_1, \ldots, Y_j, X_1, \ldots, X_k$ are sets of expressions I also write '$Y_1, \ldots, Y_j, y_1, \ldots, y_m \vdash X_1, \ldots, X_k, x_1, \ldots, x_n$' for '$Y_1 \cup \ldots \cup Y_j \cup \{y_1, \ldots, y_m\} \vdash X_1 \cup \ldots \cup X_k \cup \{x_1, \ldots, x_n\}$' unless confusion may arise. Also, I shall generally omit indication of the empty set, writing e.g. '$\vdash x_1, \ldots, x_n$' or '$X_1, \ldots, X_k \vdash$' for '$\varnothing \vdash x_1, \ldots, x_n$' and '$X_1, \ldots, X_k \vdash \varnothing$', respectively.

valid conclusions from valid premisses—at least relative to canonical frames with ideal admissibility relations. This follows directly from Fact 2.1.2. However, such arguments are only validity-preserving, not satisfaction-preserving, and this latter notion is of major relevance for an appraisal of the logic determined by a canonical frame.

Before looking at this point more closely, I shall now record a general fact that will help clarify just how "classical" the generalized logic is. To this end, let us first extend our terminology. Where $X \subseteq \mathit{EXP}_{\mathcal{L}}$ and $R \subseteq \mathit{MOD}_{\mathcal{L}} \times \mathit{MOD}_{\mathcal{L}}$, let a model $\mathcal{M}'$ count as an *(X,R)-constriction* of a model $\mathcal{M}$ (written $\mathcal{M}' \preccurlyeq^X_R \mathcal{M}$) if and only if $\mathcal{M}'$ is a simultaneous (x,R)-constriction of $\mathcal{M}$ for each $x \in X$, i.e., if and only if it is a restriction of $\mathcal{M}$ that is $\sqsubseteq$-maximal in the class $\bigcap\{\mathit{CONS}_{\mathcal{L}}(x): x \in X\} \cap R[\mathcal{M}]$; and let $\mathcal{M}'$ count as an *(X,R)-completion* of $\mathcal{M}$ (written $\mathcal{M}' \succcurlyeq^X_R \mathcal{M}$) if and only if it is a simultaneous (x,R)-completion of each $x \in X$, i.e., if and only if $\mathcal{M}'$ is an extension of $\mathcal{M}$ that is $\sqsubseteq$-minimal in the class $\bigcap\{\mathit{COMP}_{\mathcal{L}}(x): x \in X\} \cap R[\mathcal{M}]$. Using this terminology, we may distinguish two senses in which a set $X \subseteq \mathit{EXP}_{\mathcal{L}}$ can be satisfied in a model $\mathcal{M} \in K$ by a context $c \in C_{\mathcal{M}}$ (relative to a frame $\mathcal{F}$). On the one hand, we can say that X is $\mathcal{F}$-satisfied by c in $\mathcal{M}$ if and only if each $x \in X$ is $\mathcal{F}$-satisfied by c in $\mathcal{M}$—i.e., if and only if, for each $x \in X$, there is some (x,R)-constriction $\mathcal{M}''$ of $\mathcal{M}$ such that x is $\mathcal{F}$-satisfied by c in every (x,R)-completion $\mathcal{M}'$ of $\mathcal{M}''$. (This is the obvious notion, given the definitions of section 1.3.5.) On the other hand, we may say that X is *firmly* $\mathcal{F}$-satisfied by c in $\mathcal{M}$ if and only there is some (X,R)-constriction $\mathcal{M}''$ of $\mathcal{M}$ such that, for each $x \in X$, x is $\mathcal{F}$-satisfied by c in every (X,R)-completion $\mathcal{M}'$ of $\mathcal{M}''$. Schematically:

(a) $\mathit{SAT}_{\mathcal{F}}(X) = \{\langle \mathcal{M},c\rangle: \mathcal{M} \Vdash^c_{\mathcal{F}} x \text{ for all } x \in X\}$
(b) $\mathit{F\text{-}SAT}_{\mathcal{F}}(X) = \{\langle \mathcal{M},c\rangle: \langle \mathcal{M}',c\rangle \in \mathit{SAT}_{\mathcal{F}}(X) \text{ for some } \mathcal{M}'' \preccurlyeq^X_R \mathcal{M} \text{ and every } \mathcal{M}' \succcurlyeq^X_R \mathcal{M}''\}$.

In a similar fashion, we may distinguish between a set of expressions X being *$\mathcal{F}$-refuted* or *firmly $\mathcal{F}$-refuted* in a model $\mathcal{M} \in K$ by a context $c \in C_{\mathcal{M}}$:

(c) $\mathit{REF}_{\mathcal{F}}(X) = \{\langle \mathcal{M},c\rangle: \mathcal{M} \Vdash^c_{\mathcal{F}} x \text{ for no } x \in X\}$
(d) $\mathit{F\text{-}REF}_{\mathcal{F}}(X) = \{\langle \mathcal{M},c\rangle: \langle \mathcal{M}',c\rangle \in \mathit{REF}_{\mathcal{F}}(X) \text{ for every } \mathcal{M}'' \preccurlyeq^X_R \mathcal{M} \text{ and some } \mathcal{M}' \succcurlyeq^X_R \mathcal{M}''\}$.

The first sense of satisfaction and refutation yields the basic notion of $\mathcal{F}$-entailment as defined in 1.3.5(f), which can now be formulated more succinctly as follows:

(e) X is $\mathcal{F}$-entailed by Y iff $SAT_{\mathcal{F}}(Y) \cap REF_{\mathcal{F}}(X) = \varnothing$.

And the second sense yields a corresponding notion of *firm entailment*:

(f) X is *firmly* $\mathcal{F}$-entailed by Y iff $F\text{-}SAT_{\mathcal{F}}(Y) \cap F\text{-}REF_{\mathcal{F}}(X) = \varnothing$.

If a set of expressions X is firmly $\mathcal{F}$-entailed by a set Y, I will say that the argument $Y \vdash X$ is *firmly* $\mathcal{F}$-valid.

Now, it is obvious that every $\mathcal{F}$-valid argument is firmly $\mathcal{F}$-valid. However, the converse need not hold. For instance, it is easy to check that all four examples (18)–(21) are firmly $\mathcal{F}$-valid (for $\mathcal{F}$ as specified), in spite of their not being $\mathcal{F}$-valid. Keeping this in mind, a general characterization of the relationship between standard and generalized entailment can be stated as follows.

2.2.2. FACT

Let $\mathcal{L}$ be any language and R an ideal admissibility relation on $MOD_{\mathcal{L}}$. If $\mathcal{F} = \langle K, R, t, \delta \rangle$ is a canonical frame for $\mathcal{L}$ and $\mathcal{F}' = \langle K', R', t, \delta' \rangle$, where $K' = K \cap SHRP_{\mathcal{L}}$, $R' = R \upharpoonright K'$ and $\delta' = \delta \upharpoonright K'$, then:

(a) every $\mathcal{F}$-valid argument is $\mathcal{F}'$-valid; indeed,
(b) for all $X, Y \subseteq \mathbf{E}_t$, X is $\mathcal{F}'$-valid iff $Y \vdash X$ is firmly $\mathcal{F}$-valid.

Proof. (a) is obvious from the inclusion $SHRP_{\mathcal{L}} \subseteq MOD_{\mathcal{L}}$, since for every $x \in EXP_{\mathcal{L}}$ and every $\mathcal{M}' \in K'$ we have $\mathcal{V}'_R(x) = \mathcal{V}'_{R'}(x)$ and $\delta'_{\mathcal{M}'} = \delta_{\mathcal{M}'}$. As for (b), the implication from right to left (sufficiency) reduces to (a) upon noticing that firm $\mathcal{F}'$-validity coincides with $\mathcal{F}'$-validity. This is because $\mathcal{M} \in SHRP_{\mathcal{L}}$ implies $\{\mathcal{M}' : \mathcal{M}' \preccurlyeq^Z_R \mathcal{M}\} = \{\mathcal{M}' : \mathcal{M}' \succcurlyeq^Z_R \mathcal{M}\} = \{\mathcal{M}\}$ for all $Z \subseteq EXP_{\mathcal{L}}$, hence $F\text{-}SAT_{\mathcal{F}'}(Z) = SAT_{\mathcal{F}'}(Z)$ and $F\text{-}REF_{\mathcal{F}'}(Z) = REF_{\mathcal{F}'}(Z)$. Consider then the converse implication (necessity). Fix $X, Y \subseteq \mathbf{E}_t$ and assume that $Y \vdash X$ is not firmly $\mathcal{F}$-valid. Then there must exist $\mathcal{M} \in K$ and $c \in C_{\mathcal{M}}$ such that $\langle \mathcal{M}, c \rangle \in F\text{-}SAT_{\mathcal{F}}(Y) \cap F\text{-}REF_{\mathcal{F}}(X)$. This means that we can pick some $\mathcal{M}'' \preccurlyeq^Y_R \mathcal{M}$ such that $\langle \mathcal{M}', c \rangle \in SAT_{\mathcal{F}}(Y)$ for every $\mathcal{M}' \succcurlyeq^Y_R \mathcal{M}''$, while for every $\mathcal{M}'' \preccurlyeq^X_R \mathcal{M}$ there is some $\mathcal{M}' \succcurlyeq^X_R \mathcal{M}''$ such that

$\langle \mathcal{M}',c\rangle \in REF_{\mathcal{F}}(X)$. Since R is assumed to be ideal, by arguing as in the proof of Fact 2.1.2 we can find a full constriction $\mathcal{M}^{**}$ of $\mathcal{M}''$ such that $\mathcal{V}^{**}_{R,c}(y)=\mathcal{V}''_{R,c}(y)$ for every $y\in Y$. This implies that $\langle \mathcal{M}^{**},c\rangle \in SAT_{\mathcal{F}}(Y)$. Moreover, for every (X,R)-constriction $\mathcal{M}'$ of $\mathcal{M}$ we have $\mathcal{M}^{**}\sqsubseteq\mathcal{M}'$, and therefore $\mathcal{V}^{**}_{R,c}(x)\subseteq\mathcal{V}'_{R,c}(x)$ for all $x\in X$ (by monotonicity), which implies $\langle \mathcal{M}^{**},c\rangle \in F\text{-}REF_{\mathcal{F}}(X)$. Thus, $\langle \mathcal{M}^{**},c\rangle \in SAT_{\mathcal{F}}(Y)\cap F\text{-}REF_{\mathcal{F}}(X)$. By a similar argument, we can now find a full completion $\mathcal{M}^{*}$ of $\mathcal{M}^{**}$ such that $\mathcal{V}^{*}_{R,c}(x)=\mathcal{V}^{**}_{R,c}(x)$ and $\mathcal{V}^{*}_{R,c}(y)\supseteq\mathcal{V}^{**}_{R,c}(y)$ for each $x\in X$ and each $y\in Y$, hence such that $\langle \mathcal{M}^{*},c\rangle \in SAT_{\mathcal{F}}(Y)\cap REF_{\mathcal{F}}(X)$ by definition. But such a model $\mathcal{M}^{*}$ will be consistent and complete, hence sharp, hence an element of K'. Since $R'=R\upharpoonright K'$ and $\delta'=\delta\upharpoonright K'$, we can therefore conclude that the argument $Y\vdash X$ is not $\mathcal{F}'$-valid. The desired result follows then by contraposition and generalization.

2.2.3. REMARKS

Perhaps Fact 2.2.2 does not say much in this form. It says in what sense standard argument validity can be preserved in the context of non-standard models (namely, as firm validity), but it is silent on what standard valid arguments remain valid in that context. In effect, a nice conjecture would be that although the valid arguments determined by a given canonical frame may not coincide with the valid arguments of its sharp fragment, they generally form a *specifiable* subclass thereof. However, it is not quite clear how this could be made more precise.

The difficulty does not only arise in the general case, where the desired result is naturally hindered by the virtually unlimited variable composition of a canonical frame as we defined it. The point is that it seems impossible to specify the relevant class of valid arguments even if we confine ourselves to canonical frames that satisfy certain basic and rather natural closure conditions.

Here is an example. Suppose we focus on a canonical frame whose admissibility relation R is ideal and whose class of admissible models, K, is nicely closed under both operations of infimum $\sqcap$ and supremum $\sqcup$ (relative to the ordering $\sqsubseteq$). (This implies that all models in K are pairwise commensurate—e.g., a class of extensional sentential models.) With respect to such *regular* frames, as I shall call them, it is easy to establish some interesting connections between general satisfiability and sharp satisfiability, i.e., satisfiability relative to sharp models. For instance, let us say that a set of expressions X is *strongly $\mathcal{F}$-unsatisfiable* just in case it contains some $\mathcal{F}$-unsatisfiable element, and let us say that X is *strongly $\mathcal{F}$-irrefutable* just in case it contains some $\mathcal{F}$-irrefutable element. Then, if $\mathcal{F}$ is a regular frame and $\mathcal{F}'$ is its sharp fragment (as in 2.2.2), the following equivalences are easily verified:

(a) X is $\mathcal{F}$-unsatisfiable iff X is strongly $\mathcal{F}'$-unsatisfiable
(b) X is $\mathcal{F}$-irrefutable iff X is strongly $\mathcal{F}'$-irrefutable.

For, on the one hand, if X is strongly $\mathcal{F}'$-unsatisfiable (irrefutable), then it is $\mathcal{F}$-unsatisfiable (irrefutable) by Fact 2.2.2(b). On the other hand, if X is not strongly $\mathcal{F}'$-unsatisfiable, then a K-model $\mathcal{F}$-satisfying X can be obtained by taking the $\sqsubseteq$-supremum of any X-termed sequence of K'-models each of which satisfies the corresponding $x \in X$, whereas if X is not strongly $\mathcal{F}'$-irrefutable, then a K-model refuting X can be obtained by taking the $\sqsubseteq$-infimum of any sequence of K'-models each of which refutes the corresponding $x \in X$:

(c) if $\mathcal{M} = \bigsqcup\{\mathcal{M}_x : x \in X\}$ and $\langle \mathcal{M}_x, c \rangle \in SAT_{\mathcal{F}'}(x)$ for all $x \in X$, then $\langle \mathcal{M}, c \rangle \in SAT_{\mathcal{F}}(X)$
(d) if $\mathcal{M} = \bigsqcap\{\mathcal{M}_x : x \in X\}$ and $\langle \mathcal{M}_x, c \rangle \in REF_{\mathcal{F}'}(x)$ for all $x \in X$, then $\langle \mathcal{M}, c \rangle \in REF_{\mathcal{F}}(X)$.

This fact has some interest, especially insofar as regular frames arise naturally in the semantics of familiar languages (e.g., sentential languages). However, there is not much more to it. For although in classical contexts there is a very crucial link between unsatisfiability (or irrefutability) and argument validity, the link vanishes in the general case. Thus, although in classical sentential logic the following holds for all sets of sentences X and Y:

(e) X is $\mathcal{F}'$-entailed by Y iff $Y \cup \{\neg x : x \in X\}$ is $\mathcal{F}'$-unsatisfiable iff $X \cup \{\neg y : y \in Y\}$ is $\mathcal{F}'$-irrefutable,

in the presence of incomplete or inconsistent models these equivalences may fail, as one can easily check by considering examples (18)–(21).

Even with respect to regular frames, then, and even with reference to languages of a very simple sort—such as sentential languages—the characterization of argument validity seems to run afoul of standard patterns. My conjecture is that in most cases the valid arguments do not in fact form a specifiable subclass of the arguments rated valid in the corresponding sharp fragment. And the reason I offer is that the notion of logical consequence becomes extremely language sensitive in the presence of unsharp models. We shall see below some examples. On an intuitive understanding of the notion of "substitution instance", the general point can be put thus: given an $\mathcal{F}$-valid argument $Y \vdash X$, there generally exist $Y'=\{y': y\in Y\}$ and $X'=\{x': x\in X\}$ such that (i) each $z'\in Y'\cup X'$ is a substitution instance of the corresponding $z\in Y\cup X$, and (ii) $Y' \vdash X'$ is not $\mathcal{F}$-valid. (Of course, this is not the case with every $\mathcal{F}$-valid argument: we have already remarked that the structural principles of reflexivity, reiteration, cut, and thinning hold unrestrictedly.)

With all this, some consequences of Fact 2.2.2 are nonetheless instructive. For instance, let $\mathcal{F}=\langle K, R, t, \delta\rangle$ be as stated. It is apparent that every argument of the form

$$y \vdash x,$$

and more generally every argument with at most one premiss and at most one conclusion, is $\mathcal{F}$-valid just in case it is $\mathcal{F}'$-valid. This is because the contrast between $\mathcal{F}$-satisfaction (refutation) and firm $\mathcal{F}$-satisfaction (refutation) only arises in the presence of sets with multiple expressions. This also implies that Fact 2.1.2 follows—as it should—as a special case of Fact 2.2.2 where $Y=\varnothing$. Moreover, it is apparent that if we only consider consistent models, then the notions of $\mathcal{F}$-satisfaction and firm $\mathcal{F}$-satisfaction coincide—i.e., a consistent model $\mathcal{M}$ firmly $\mathcal{F}$-satisfies a set of sentences Z if and only if it $\mathcal{F}$-satisfies Z. Thus, in that case an argument of the form

$$Y \vdash x$$

is $\mathcal{F}$-valid just in case it is $\mathcal{F}'$-valid. Likewise, $\mathcal{F}$-refutation and firm $\mathcal{F}$-refutation coincide on complete models. Hence, if we only consider such models, an argument

$$y \vdash X$$

is $\mathcal{F}$-valid just in case it is $\mathcal{F}'$-valid. We can sum up all of these consequences as follows:

(f) if $X \lesssim 1$ and $Y \lesssim 1$, then an argument $Y \vdash X$ is $\mathcal{F}$-valid iff it is $\mathcal{F}'$-valid;

(g) if $X \lesssim 1$ and $K \subseteq \bigcap\{CONS_{\mathcal{L}}(y): y \in Y\}$, then an argument $Y \vdash X$ is $\mathcal{F}$-valid iff it is $\mathcal{F}'$-valid;

(h) if $Y \lesssim 1$ and $K \subseteq \bigcap\{COMP_{\mathcal{L}}(x): x \in X\}$, then an argument $Y \vdash X$ is $\mathcal{F}$-valid iff it is $\mathcal{F}'$-valid.

The picture is best appreciated if we observe that the opposition between $\mathcal{F}$-validity and firm $\mathcal{F}$-validity is essentially a matter of distributive versus collective reading of an argument.[72] To say that an argument is $\mathcal{F}$-valid is to say that whenever each one of its premisses is satisfied, so is some one of its conclusions. To say that the argument is firmly $\mathcal{F}$-valid is to say that whenever the "conjunction" of its premisses is satisfied, then so is the "disjunction" of its conclusions. To make this more explicit, let us focus again on our typical examples—sentential and elementary logic. Let us define collective satisfaction and refutation as follows, where $X \subseteq \mathbf{E}_0$:

(i) $C\text{-}SAT_{\mathcal{F}}(X) = \bigcap\{SAT_{\mathcal{F}}(x_1 \wedge \ldots \wedge x_n): n>0 \text{ and } \{x_1,\ldots,x_n\} \subseteq X\}$

(j) $C\text{-}REF_{\mathcal{F}}(X) = \bigcap\{REF_{\mathcal{F}}(x_1 \vee \ldots \vee x_n): n>0 \text{ and } \{x_1,\ldots,x_n\} \subseteq X\}$.

(In case $n=1$, it is understood that $x_1 \wedge \ldots \wedge x_n = x_1 \vee \ldots \vee x_n = x_1$.) Accordingly, let us introduce one more notion of entailment between sets of sentences

(k) X is *collectively* $\mathcal{F}$-entailed by Y iff $C\text{-}SAT_{\mathcal{F}}(Y) \cap C\text{-}REF_{\mathcal{F}}(X) = \varnothing$,

As before, this notion of entailment induces a corresponding property of *collective $\mathcal{F}$-validity* applying to arguments. We then have the following corollaries of Fact 2.2.2.

2.2.4. COROLLARIES

Let $\mathcal{L}$ be any language and let $\mathcal{F} = \langle K, R, t, \delta \rangle$ be a canonical frame for $\mathcal{L}$, where R is ideal, $t=0$, and $\delta_{\mathcal{M}}=1$ for all $\mathcal{M} \in K$. Then:

---

[72] The terminology is from Rescher [1979], §5, where a similar point is made. See also Rescher & Brandom [1980], §5 and *passim*, and the other works cited in notes 52 and 53 above.

(A) If $\mathcal{L}$ is a sentential language with a binary connective $\downarrow$, as in Example 1.1.4(A), and K is the class of all 2-valued, extensionally adequate sentential models for $\mathcal{L}$, as in Example 1.2.4(A), then an argument $Y \vdash X$ is collectively $\mathcal{F}$-valid iff it is rated valid in classical sentential logic.

Proof. We just observe that the equalities $C\text{-}SAT_{\mathcal{F}}(X) = F\text{-}SAT_{\mathcal{F}}(X)$ and $C\text{-}REF_{\mathcal{F}}(X) = F\text{-}REF_{\mathcal{F}}(X)$ hold for all $X \subseteq \mathbf{E}_0$. This is easily verified in view of conditions (c)–(d) in Example 1.3.4(A). The desired result then follows directly from Fact 2.2.2, considering that the $\mathcal{F}'$-valid arguments are exactly the valid arguments of classical sentential logic. Indeed, if we confine ourselves to arguments with finite sets of premisses and conclusions, the corollary reduces to a straightforward instantiation of remark (f) in the previous section.

Remarks. Corollary 2.1.4(A) can be seen as a special case of the present corollary, taking $Y=\varnothing$: tautologies are just collectively understood conclusion sets in premiss-free arguments (or, what amounts to the same thing, tautologies are entailed by any set of premisses). Likewise, the standard notion of a contradiction as a sentence entailing any set of conclusions (or, equivalently, as a collectively understood premiss set of a conclusion-free argument) is given by the special case $X=\varnothing$. The interesting thing is that now the notion of entailment allows us to go beyond such special cases and *therefore* beyond the prospect of classical semantics: while it is true that tautologies or contradictions such as (1) or (2) retain their traditional logical status, this is not the case for the corresponding arguments (1') and (2') (understood distributively):

(1) $A \vee \neg A$
(2) $A \wedge \neg A$
(1') $\vdash A, \neg A$
(2') $A, \neg A \vdash$

In fact, the latter represent a way of expressing the assumptions of completeness and consistency, respectively, that characterize the sharp models of standard bivalent semantics, in contrast with (1) and (2). It is precisely this contrast in terms of expressiveness that was pointed out in our earlier discussion of sentence validity, and it is this contrast that gives content to the distinction between collective and distributive argument validity. Indeed, it is precisely the non-validity of such arguments as (1') and (2') that prevents gaps and gluts from spreading around in the entire canonical frame $\mathcal{F}$. Contradictions explode consequentially, just as in classical logic, and tautologies implode. However, accepting unsatisfiable premiss sets or irrefutable conclusion sets is generally not tantamount to endorsing unsatisfiable (contradictory) premisses or irrefutable (tautologous) conclusions, and this protects $\mathcal{F}$ from the danger of logical trivialization.[73]

There are many other argument forms in which this becomes manifest. For instance, it is immediately seen that although arguments (22) and (23) are $\mathcal{F}$-valid, the related classical principles (24) and (25) may fail (for instance, when $B = \neg A$ and A is either undetermined or overdetermined, respectively):

(22) $A \wedge B \vdash A$
(23) $A \vdash A \vee B$
(24) $A, B \vdash A \wedge B$
(25) $A \vee B \vdash A, B$.

This reflects the partiality of conditions (c)–(d) in Example 1.3.4(A) (which specify the semantic behavior of '$\wedge$' and '$\vee$') and makes $\mathcal{F}$ a non-adjunctive, non-subjunctive canonical frame.[74]

Another important classical principle that fails to hold unrestrictedly is disjunctive syllogism (or also *modus ponens*) as in (26) below: this argument fails whenever A is both true and false and B is at most false (not true); the dual argument (27) fails in a like manner whenever A is undetermined and B is at least true (possibly also false):

(26) $A, \neg A \vee B \vdash B$
(27) $B \vdash A, \neg A \wedge B$.

[73] In this sense, the resulting logic is weakly non-Scotian, in the terminology of Marconi [1981].

[74] Non-adjunctive logics have been pioneered by Jaśkowski [1948]. Some authors regard the failure of principle (24) as a sign that the symbol '$\wedge$' has departed from its normal interpretation, i.e., from meaning 'and'. See e.g. Priest & Routley [1989a], fn. 159, where this departure is compared with that of intuitionistic negation. See also Priest & Routley [1989b], § 2.1. Analogous remarks would presumably apply to disjunction's failing to satisfy principle (25).

Of course, certain classes of arguments remain valid with respect to certain distinguished families of non-classical sentential models. For instance, in view of what was said in section 2.2.3 above, point (g), it is clear that all classically valid arguments with a single conclusion remain valid if we confine ourselves to the class of all consistent (possibly incomplete) models in K. Thus, within such limits the classical sentential calculus is complete not only with regard to sentence validity, as assured by Corollary 2.1.4(A), but also with regard to entailment.[75] In general, however, things do not work that way and the connection with classical logic becomes blurry.

Now, the point of looking at collective validity is precisely that it provides a way of expressing the logical prospect that validates all classically valid arguments with regard to all K-models. Collective validity is by definition disrespectful of any notion of distributive satisfaction or refutation, and in this sense it sheds light on the equivalence between firm and collective validity as opposed to validity in general. Thus, all of the above arguments—more generally, all valid arguments of classical sentential logic—will continue to hold unrestrictedly if we consider firm or collective validity in place of validity *tout court*. The reason is, quite naturally, that firm validity defines a link between an inconsistent or incomplete model and its sharp constrictions/completions which eventually *conceals* the peculiarity of the former in favor of the normality of the latter: firm validity leaves no room for unsatisfiability without contradiction, no room for irrefutability without tautologousness. This is sufficient to restore a systematic connection with classical logic, and the notion of collective validity just expresses this in more explicit terms.

It is also worth observing that all argument forms considered above share an interesting feature, in that they all involve a non-contingent premiss set or conclusion set (where a set of expressions X qualifies as *$\mathcal{F}$-contingent* when X is neither $\mathcal{F}$-unsatisfiable nor $\mathcal{F}$-irrefutable). More precisely, call an argument $Y \vdash X$ *strongly $\mathcal{F}$-valid* if and only if there exist $Y' \subseteq Y$ and $X' \subseteq X$ such that (i) $Y' \vdash X'$ is $\mathcal{F}$-valid, and (ii) if every $y \in Y'$ is $\mathcal{F}$-satisfiable and every $x \in X'$ is $\mathcal{F}$-refutable, then Y' and X' are

[75] This result is familiar from supervaluational semantics. See e.g. van Fraassen [1970b], p. 96.

both $\mathcal{F}$-contingent. Then the above examples can all be explained in the same way: the arguments fail to be $\mathcal{F}$-valid because they are not strongly $\mathcal{F}'$-valid. This should not, however, be taken as a specification of the class of $\mathcal{F}'$-valid arguments that remain $\mathcal{F}$-valid, for there are arguments whose $\mathcal{F}$-validity depends crucially on involving non-$\mathcal{F}'$-contingent sets of premisses or conclusions. For instance, suppose R is the *universal* admissibility relation $K \times K$, so that every sharpening counts as admissible. Then the argument

(28) $p \leftrightarrow q \vdash p, \neg p.$

(where $p,q \in SYM_{\mathcal{L}}$) is a case in point. This argument is $\mathcal{F}$-valid, because whenever the biconditional $p \leftrightarrow q$ is at least true, either p (and q) or $\neg p$ (and $\neg q$) must be at least true as well. Yet (28) is not strongly $\mathcal{F}'$-valid.

Above I spoke of the language sensitivity of the general notion of $\mathcal{F}$-entailment. What I mean by that emerges forcefully from these examples. If $R = K \times K$, then q is semantically independent from p, and that is enough to guarantee the $\mathcal{F}$-validity of (28) as well as, say, (29) or (30) as opposed to (29') and (30'):

(29) $p, q \vdash p \wedge q$

(30) $p \vee q \vdash p, q$

(29') $p, \neg p \vdash p \wedge \neg p$

(30') $p \vee \neg p \vdash p, \neg p.$

Nor is this a peculiarity of the specific notion of $\mathcal{F}$-validity with which we are working: similar considerations would apply if we were working for instance with the relation of negative $\mathcal{F}$-entailment $\vDash_{\mathcal{F}}^{-}$, or with the double-barreled relation $\vDash_{\mathcal{F}}^{\pm}$.

The moral of all this is a rather negative characterization of the logic of $\mathcal{L}$. We know from 2.1.4 that an axiomatization is available—e.g., the classical sentential calculus—that allows us to derive as theorems all and only those $\mathcal{L}$-sentences that are $\mathcal{F}$-valid. However, this is no axiomatization of validity as applied to arguments, except with respect to certain classes of models. The present corollary shows that not everything is lost. But if we take the phenomenon of language sensitivity at face value, then the foregoing remarks suggest that the loss is nevertheless severe: the set of valid arguments is likely to be not even recursively enumerable.

I shall now conclude with some remarks on the logic of elementary languages.

(B) If $\mathcal{L}$ is an elementary language with a set $V_{\mathcal{L}}$ of name variables and a binary quantifier $\langle\downarrow,v\rangle$ for each $v\in V_{\mathcal{L}}$, as in Example 1.1.4(B), and if K is the class of all 2-valued, non-empty, logically adequate elementary models for $\mathcal{L}$, as in Example 1.2.4(B), then an argument $Y\vdash X$ is collectively $\mathcal{F}$-valid iff it is rated valid in classical elementary logic.

Proof. Again, it suffices to note that in view of 1.3.4(A)(c)–(d), the equalities $C\text{-}SAT_{\mathcal{F}}(X)=F\text{-}SAT_{\mathcal{F}}(X)$ and $C\text{-}REF_{\mathcal{F}}(X)=F\text{-}REF_{\mathcal{F}}(X)$ hold for all $X\subseteq\mathbf{E}_0$. Then the result follows directly from Fact 2.2.2, considering that the $\mathcal{F}'$-valid arguments are exactly the valid arguments of classical elementary logic.

Remarks. Much of what was said above about sentential logic can be extended *mutatis mutandis* to the present case. In particular, the corollary can be viewed as a generalization of 2.1.4(B), which corresponds to the special case $Y=\varnothing$. It is, however, interesting to observe how the delicate interplay between sentence validity and argument validity—an interplay that makes us go beyond such special cases and consequently beyond the prospect of classical semantics—becomes more intricate in the context of elementary languages.

Consider again the paradigmatic cases discussed earlier. Although such sentences as

(5) $Px\vee\neg\bigwedge vPv$

(6) $Px\wedge\neg\bigvee vPv$

retain their traditional logical status, as we saw in section 2.1.4(B), the corresponding arguments do not:

(5") $\vdash Px,\neg\bigwedge vPv$

(6") $Px,\neg\bigvee vPv\vdash$.

(Compare (20)–(21).) In other words, (5") and (6") are $\mathcal{F}'$-valid but not $\mathcal{F}$-valid. This is because the status of (5") and (6"), unlike that of (5) and (6), does reflect the critical assumptions of completeness and consistency (for instance, relative to P) that were not included in the characterization of $\mathcal{F}$. The underlying explanation should be obvious given our earlier discussion of (1)–(2) versus (1')–(2'): we are dealing with distinctions that are indeed characteristic of our general semantic framework, and the only way to abridge them is by interpreting arguments collectively rather than distributively (i.e., in terms of firm entailment rather than entailment proper). In addition, however, we can now also consider inferential principles such as the following:

(5''') $\bigwedge vPv \vdash Px$
(6''') $Px \vdash \bigvee vPv$

These principles are very closely related to (5'') and (6''). And they *are* $\mathcal{F}$-valid: for (5''') and (6''') are $\mathcal{F}'$-valid, hence collectively $\mathcal{F}$-valid, and we have seen that for arguments with a single premiss and a single conclusion, firm or collective $\mathcal{F}$-validity reduces to $\mathcal{F}$-validity.

Now, how exactly are these patterns of validity related? One answer is that (5''') and (6''') express essentially the same principles as are reflected in the logical status of (5) and (6), respectively, and must be distinguished from (5'') and (6'') essentially for the same reasons. In effect, it follows from our Corollary that $\mathcal{F}$ supports the following general equivalences (a form of "deduction theorem" for singular arguments) for all $A,B \in \mathbf{E}_0$:

(a) $A \vdash B$ is $\mathcal{F}$-valid iff $\vdash \neg A \vee B$ is $\mathcal{F}$-valid iff $A \wedge \neg B \vdash$ is $\mathcal{F}$-valid.

Thus, the correspondence between (5)–(6) and (5''')–(6''') can be seen as instantiating a more general feature of $\mathcal{F}$. These principles do not reflect any general semantic constraint on the range of admissible interpretations. Rather, they express certain fundamental principles of inference whose validity—relative to a canonical frame based on an ideal admissibility relation—is not questioned by the abandonment of the semantic requirements of completeness and consistency.

There is a second, subtler answer. It is that (5''') and (6''') express a pattern of reasoning which lies behind the logical status of (5) and (6), respectively, and which reflects the semantic interpretation of the quantifiers relative to $\mathcal{F}$. Sentence (5) is always true *because* it is not possible to satisfy a universal statement '$\bigwedge vPv$' while at the same time refuting a corresponding instance 'Px'; sentence (6) is always false because it is not possible to satisfy a singular statement 'Px' while at the same time refuting the corresponding existential quantification '$\bigvee vPv$'. In neither case do the assumptions expressed by (5") and (6") play any role; and in both cases they can in fact be rejected.

There is one more point, however. We have seen above (Corollary 2.1.4(B)) that the intuitive status of (5) and (6) may change depending on whether we take the relevant source of incompleteness or inconsistency to lie in the predicate P or in the name x. With (5''') and (6''') the situation is not different. The account given here fits well with the first option. But when P is sharp and the gap or glut lies in the referential pattern of x, then (5''') and (6''') are open to various interpretations. Take (5'''), for instance, and consider a model $\mathcal{M} \in K$ and an assignment $c \in C_{\mathcal{M}}$ where x is a constant lacking a denotation. We have seen that if R is the ideal admissibility relation, every (x,R)-completion of $\mathcal{M}$ is bound to have the same domain of quantification as $\mathcal{M}$ itself, which implies that x can only be made to denote an element already available in $\mathbf{D}_1$. Thus, if $\bigwedge vPv$ is satisfied by c in $\mathcal{M}$, Px must be satisfied too. This is why (5''') comes out $\mathcal{F}$-valid, precisely as (5) did. However, we have also seen that there is an alternative way of understanding the denotationlessness of a name x. From this alternative point of view, x may lack a denotation because it may fail to refer to an entity available in the domain of quantification ('Pegasus'). Hence the validity of (5''') may be found unacceptable insofar as it involves an existential presupposition with respect to the denotation of the name x, just as the traditional principle of *reductio ad subalternatam*

$$\text{(31)} \quad \bigwedge v(Pv \rightarrow Qv) \vdash \bigvee v(Pv \wedge Qv)$$

is faulty of existential presupposition with respect to the denotation of the predicate P.[76] This intuition cannot be accommodated within the canonical frame $\mathcal{F}$. But it becomes perfectly legitimate if we refer to other (possibly non-canonical) frames. In particular, we may consider the frame resulting from $\mathcal{F}$ by replacing its ideal admissibility relation by a (strongly) free variant, as in 2.1.4(B). Then an x-completion of $\mathcal{M}$ would be allowed to interpret x in c by means of a denotation *foreign* to the domain of quantification. And that would block the validity of (5'''); for an assignment that satisfied the premiss $\bigwedge vPv$ in an incomplete

---

[76] This was the point of Leonard [1956], which eventually led to the development of so-called free logics (see above, note 16).

model might refute (i.e., leave undefined) the conclusion Px insofar as it would refute (i.e., falsify) it in some of the model's completions.

As before, this alternative account would be in the spirit of free logic. But there is a peculiarity. While the link between (5) and (6) is preserved in the resulting semantics (these sentences fall together when x fails to denote), the link between (5''') and (6''') could be lost. For (6''') *could* be valid even if $\mathcal{F}$ were modified as indicated. For instance, suppose the relevant admissibility relation is universal, $R = K \times K$. Then an atomic sentence such as Px can be satisfied only in a model $\mathcal{M} \in K$ and by an assignment $c \in C_{\mathcal{M}}$ such that $\mathcal{V}_c(x) \subseteq \mathbf{D}_1$. Otherwise Px would be undefined, since we could extend (a restriction of) $\mathcal{M}$ by interpreting x so as to satisfy Px as well as to refute Px. Hence, if $\mathcal{M}$ is such a model, it will also satisfy $\bigvee vPv$ regardless of whether $\mathcal{M}$'s sharpenings are defined via R or via one of its free variants. (Compare this with the validity of (28).)

Among other things, this means that the logic induced by a free counterpart of $\mathcal{F}$, though closer in spirit to a free logic, is certainly not fully axiomatizable by a free-logic calculus. That is, ordinary systems of free logic cannot be argument complete with respect to the semantics in question. For such systems are sound with respect to the general patterns of validity given in (a) above, whereas the theory at issue would violate them: argument (6''') would be valid but the disjunction

(32) $\neg Px \vee \bigvee vPv$

would not. (It would fail just as (5) does.)

Another way of describing this state of affairs is the following. Suppose K is restricted in the usual way so as to interpret a binary predicate $\equiv$ as identity. Then we have seen that a sentence of the form

(7) $\bigvee v(x \equiv v)$

may be taken to assert that x denotes an element of the given domain of quantification (i.e., that x actually exists[77]) and it is natural to look at the

[77] This is in the spirit of Quine's [1939] dictum that "to be is to be the value of a bound variable". This reading of (7), and its use in (33)–(34) below, goes back to Leblanc & Hailperin [1959] and Hintikka [1959], though it was already present in Whitehead & Russell's treatment of definite descriptions in [1910], *14.

following restricted principles as substitutes for (5''') and (6''') in free logic, respectively:[78]

(33) $\bigwedge vPv, \bigvee v(x{\equiv}v) \vdash Px$
(34) $Px, \bigvee v(x{\equiv}v) \vdash \bigvee vPv.$

However, it is easy to verify that the non-canonical frame under discussion would treat every argument of the following sort as valid:

(35) $Px \vdash \bigvee v(x{\equiv}v).$

(The reasoning parallels the explanation given above in relation to (6''').) Thus, by Thinning and Cut, (34) turns out to be valid just in case (6''') is valid—which takes us back to the point made above. Nor is there any clear way we can keep these inferential patterns under control.[79] For along with (35) the following sort of argument is also valid in the semantics at issue, and for the same reason:

(36) $\neg Px \vdash \bigvee v(x{\equiv}v).$

So, if every valid argument were provable, (35) and (36) would allow us to derive (7) from

(3) $Px \vee \neg Px$

using only sentential logic. And (7) is all but an acceptable theorem of free logic.

Such negative results indicate that even a free perspective, based on amending the characterization of $\mathcal{F}$ in 2.2.4(B) by referring to the non-canonical frame induced by a free variant of R, would not make up for the pathology of the entailment relations allowed by our general semantic framework. In fact, such a perspective is even more elusive insofar as reference to the notion of collective validity would not be of any help either: as the above examples indicate, Corollary 2.2.4(B) does not continue to hold if $\mathcal{F}$ is amended in the indicated way. There is, however, one interesting way of explaining this upshot, one that is not available in

[78] In the absence of $\equiv$ no such substitutes can be found, except for the universal closures (5') and (6') mentioned in section 2.1.4 (see Meyer, Bencivenga & Lambert [1982]).

[79] Here I follow an argument due to van Fraassen [1966a], §7.

the case of $\mathcal{F}$. We may say that the account under discussion allows for relations of *presupposition* (as opposed to *implication*) which have no syntactic counterpart.[80] Intuitively, let us say that a sentence A is presupposed by a sentence B iff satisfaction of the former is a necessary condition for the latter to receive a definite semantic value at all. More generally, where $X,Y \subseteq \mathbf{E}_0$ and $\mathcal{F}$ is any frame (canonical or not) relative to $\mathcal{L}$, let us add the following to the definitions of entailment of section 1.3.5:

> X is *$\mathcal{F}$-presupposed* by Y iff X is $\mathcal{F}$-entailed by all sets $\{A_y\colon y \in Y\}$ such that $A_y \in \{y, \neg y\}$ for all $y \in Y$.

Then the suggestion could be made that in the free, non-canonical frame under discussion there are relations of presupposition (such as (35)–(36)) which should be set apart. They can be viewed as expressing some crucial semantic relations among (sets of) sentences. But they play no role in the actual derivation of a conclusion from a given set of premisses. *Therefore* they escape the bounds of any rule-based system of deductive inference.

---

[80] This was the suggestion of van Fraassen [1968, 1969], referring to the notion of presupposition introduced by Frege [1892] and stressed by Strawson [1950, 1952] in relation to its pervasive role in ordinary language (see Soames [1989] and Beaver [1997] for overviews). For further thoughts along these lines, see van Fraassen [1970b], Herzberger [1975a, 1975c], Stiver [1975], and Sternefeld [1979], *inter alia*.

CONCLUDING REMARKS

Much still remains to be done for a proper assessment of the mathematical and philosophical ramifications of the semantic apparatus outlined above. There are some substantial conjectures that I have left open—for instance, concerning the non-characterizability of the entailment relation. And there are some basic, fundamental issues that I have barely broached—for example concerning the "modal flavor" of our valuations, or in regard to the ontological status of the entities that make up languages and models, and the corresponding notions of denotation, reference, semantic value. Even so, I hope I have covered enough aspects and worked through enough details and examples to make the approach intelligible.[81]

One thing that perhaps remains to be emphasized is that the approach itself has been guided by two distinct and somehow complementary aims. On the one hand, my purpose has been to lay some groundwork for a broad perspective on the semantic landscape. In this sense,

---

[81] Some highlights on these issues may be gathered from the relevant literature on supervaluational semantics. For instance, Herzberger [1982], Bencivenga [1983] and Woodruff [1984a, 1991] contain relevant metamathematical results (e.g., about compactness); Herzberger [1975a, 1975b, 1980], Silvestrini [1981], Martin [1984], and Brink [1987] contain useful material about the link between the supervaluational technique and other semantic options (e.g., classical or 3-valued logic); Barba Escribà [1987, 1989, 1993] has investigated the modal flavor of supervaluational semantics for certain types of frames; and Sorensen [1988], Dummett [1995], and Collins & Varzi [1997] have addressed the question of whether and to what extent an incomplete model can be sharpened in actual cases. (As Dummett [1991], p. 74, has pointed out, in relation to specific languages the hard work will come precisely in laying down which systems of sharpenings—if any—are to count as admissible.) More philosophical material, concerning for instance the bearing of supervaluationism on various questions in metaphysics and the philosophy of language, may be found in Rasmussen [1990] and Burgess [1997].

the resulting picture is more properly described, not as a semantic theory, but as a general semantic *framework* encompassing a large variety of specific, possibly incompatible theories: as the form of the language and the admissibility conditions on the models are varied, the theory varies. On the other hand, if we take formal semantics to be a contribution to the formal theory of meaning, then the account offered here can also be described as a semantic *theory*, though of a most abstract, general, and language-independent sort. It does not say anything about what the meaning of particular expressions is; but it says what—in general—the semantic value of any expression is like.

In pursuing these objectives I have sought to allow for a most general and liberal attitude, both with respect to the relevant notion of language and with regard to the sort of structures that can be used to interpret a given language. In particular, I have advocated a tolerant perspective on certain phenomena of semantic incompleteness and inconsistency that have been ostracized from the realm of classical logical semantics, but whose importance has nonetheless been recognized in various fields and for various purposes. Roughly put, the argument is that although one may assume that a language's expressions are always intended to have a definite semantics, there is no *a priori* guarantee that the underlying conditions will always be completely and consistently fulfilled. And since there is no general antidote against semantic incompleteness and inconsistency, there is no general and effective way that incompletenesses and inconsistencies can be ruled out without ruling out a great deal of unproblematic cases as well. A general semantic framework must therefore be capable of handling such phenomena—it must be capable of accounting for the main semantic notions independently of any specific constraints that one may want to impose upon the class of admissible models. This is not to deny that inconsistency and incompleteness have a negative flavor. They do, or so one could argue. But they are not for that reason to be banished from the realm of semantics. To put it into a slogan, consistency and completeness are two important *desiderata* for the purpose of semantic systematization—but neither is a *necessitatum*.[82]

---

[82] Adapted from Rescher and Brandom [1980], p. 137.

This was the general philosophy behind this work. I would now like to conclude with a couple of remarks on the actual proposal developed here, and to give a few indications of further lines of research that I think worth pursuing, but which in the foregoing could only be mentioned *en passant*.

A first remark concerns the cost-benefit analysis of the proposed framework. I believe that an important feature lies in its abstractness and in the corresponding generality and uniformity of treatment: the framework has not been formulated in response to any specific needs, or with respect to any particular languages, or for the purpose of any intended applications. It is based on some basic considerations concerning the way models and languages interact, regardless of the specific models and languages one may be interested in. I deem this to be important if our objective is to gain deeper understanding of the basic semantic notions. And there are some practical advantages too. For it follows that we do not need specific theories to be worked out afresh for each case: as I said above, all we need to do is to provide a syntactic category for the symbols we want to study (eventually along with a suitable structural operation) and then specify which, among the indefinitely many structures that provide a possible interpretation of the language, are to count as "admissible" models. Everything else is fixed once and for all. It is this feature that I regard as advantageous. Among other things, it justifies the top-down policy of reasoning with abstract structures first, and then move on to consider exemplifications and applications to concrete classes of languages such as those discussed in our examples.

On the other hand, we have also seen that the mathematical side is not very rewarding, not to say discouraging. Some fundamental notions (such as argument validity, with all that goes along with it) seem to escape the control of formal, recursively specifiable systems; and that makes things look cumbersome both theoretically and from a practical perspective. Even more so if we consider that our semantic machinery involves a complex and computationally heavy valuational process, requiring any expression to be evaluated on a variety of "potential" models before being assigned some actual semantic values (if any): if no equivalent syntactic system is available, establishing the validity of an argument by means of purely semantic methods may require a lot of work. (Note that this outcome is not simply a by-product of the tolerant setting of the framework, but depends on the specific form of the account advocated here. Other possible accounts, including some of the valuational strategies mentioned in section 1.3, would enjoy different mathematical properties at no cost in terms of generality or abstractness.[83] They would, however, reflect different views on what goes on.)

As I see it, there are two opposite ways of interpreting these negative results. If we believe that some form of recursive computability is an essential ingredient for any theory to have some cognitive plausibility at all, then there is little to be added: the account offered here surely fails to comply with this standard. (At best, we are left with the task of explaining why things are so and not otherwise.) If, on the other hand, we see semantics as a truly autonomous field of research, not concerned with implementational issues except for practical reasons, then those negative results must simply be taken at face value: things turn out to be more complicated than expected, and the achievement of the desired degree of generality is a costly matter after all. We do have a precise account of the basic semantic notions. And we do have a rigorous procedure to establish the semantic status of arbitrary expressions or sets of expressions with respect to an enormous variety of models. The drawback is: we cannot hope to keep this account under control by means of deductive rule-based systems. This is negative and, in a sense, unfortunate; but it is negative and unfortunate mostly at a practical level. It affects the easiness of applicability of the framework, not its theoretical status. Theoretically, it simply has the effect of making the account depend even more heavily on its ultimate assumptions: the set-theoretical ontology on which the entire framework has been developed, and the corresponding set-theoretical principles that have been taken for granted throughout the development (see the Appendix). I see nothing wrong with this.

This issue ties in with a second remark that I wish to make. As I already mentioned, one important feature of the proposed account is that different semantic theories can be seen as the result of selecting different languages and/or different classes of models and an admissibility relation to go along with them. Thus, for instance, the difference between a

---

[83] See the case discussed in Varzi [1994c].

classical two-valued semantics and a many-valued one is simply a matter of choosing the right models, just as the difference between, say, stratified versus non-stratified, or intensional versus extensional semantics. Likewise, different theories within the same language can be treated as the result of selecting different classes of models as the only admissible ones—effectively, as the result of fixing the interpretation of certain expressions. This is hardly noteworthy if by a theory we simply mean a set of sentences with a common "logical" vocabulary: in that case we are just dealing with a natural generalization of a standard model-theoretic feature.[84] What is relevant is that the same applies to theories under a much broader understanding, including when the logical vocabulary is allowed to vary along with the rest. For in that case we come up with a form of conventionalism with respect to logic itself: logic (that is, any logic) becomes neither more nor less than a theory among others. A logic becomes a theory just as, say, Peano arithmetic or the theory of partial orderings are theories, semantically speaking: they are specified by a certain set of theses expressed in a distinguished language, and these theses in turn are singled out by specifying the intended interpretation of certain distinguished terms or operations available in the language (those terms and operations that are regarded as theory-specific primitives).

Now, in a sense this kind of conventionalism is obvious. It amounts to saying that the difference between competing logical theories is essentially a matter of disagreement on the interpretation of certain logical words or operations; the way some logical terms or operations are *understood* changes as we move from one theory to another.[85] The point I am stressing is that this change of understanding can be treated in plain model-theoretic terms. To endorse a certain logic is to select a language and a suitable (canonical) frame. To endorse a different logic is to select a different canonical frame. All of our examples bear witness to this account: the framework *per se* is free of any commitment and is common to all theories.

Of course, this is not to deny that one can *define* some semantic properties in such a way as to make them invariant across models. For

---

[84] See for instance Chang & Keisler [1973].

[85] This was the gist of Quine's [1970] dictum: change of logic, change of subject.

instance, the structural properties of Reflexivity, Reiteration, Thinning, and Cut mentioned in 1.3.5 are universally valid. They are satisfied by the relation of entailment regardless of the language under examination and irrespectively of the actual size and composition of the relevant canonical frame: they simply reflect a certain way of defining entailment *in general*. So in a sense some logic does show up in the overall semantic framework. But this and similar facts are explained in terms of metalinguistic conventions. Reflexivity, Reiteration, etc. hold in every frame because these properties reflect certain logical and set-theoretical principles that we are presupposing in the very definition of entailment. On a different interpretation of, say, the connective 'if... then', or of the notion of 'set' involved in the definitions, the picture might look quite different. The prospect of conventionalism can thus be coherently maintained provided only that it is reiterated at each level of the metalinguistic hierarchy.

This view can be defended independently of any universal semantics program.[86] But it certainly comes out emphasized on the perspective examined here. For, on the one hand, we have seen that our notion of a model allows for a uniform treatment of logical words of different sort —connectives as well as operators such as quantifiers and other variable binders that one might be tempted to see as running afoul of the account.[87] There is no significant difference among them, as long as we interpret them in the right sort of contextual models. On the other hand, the tolerance of incomplete and inconsistent structures as *bona fide* models discloses a broader landscape than is allowed in standard semantic theories, accommodating a variety of otherwise intractable cases (as long as we define the relevant admissibility relation accordingly [88]). For instance, we have seen that definite descriptors—traditionally a source of difficulties from the standpoint of classical semantics—can be

---

[86] I have done so in Varzi [1994b], from which the preceding paragraph is borrowed. See also the final sections of my [1993] and [1994a]. For a recent debate, see Etchemendy [1990], Sher [1991, 1999], and García-Carpintero [1993].

[87] Work in the tradition of Cresswell [1973] is indicative of this temptation, also in view of the result of Bar-Hillel, Gaifman & Shamir [1960] mentioned in note 5 above.

[88] For connections with the semantics of supervaluations—and a contrasting view—see Bencivenga [1984].

dealt with in several ways in the presence of non-sharp models, yielding different logics depending on how exactly one gets around to do that.

So much for the general issues. The primacy of semantics emerges strongly both if we look at the picture in terms of poor axiomatizability and if we examine it in terms of characterizability of logico-syntactic notions. It remains to be seen what is gained by all this—for instance, how the framework can be exploited to deepen our understanding of such crucial concepts as truth and logical entailment; and how it can be used to accommodate such recalcitrant phenomena, typical of the language of ordinary discourse, as vagueness, incoherence, deficiency of meaning, or the paradoxes. As I said above, this is not a secondary motivation behind this work. If we see formal semantics as a tool for deepening our understanding of scientific as well as ordinary discourse, then going beyond the repertoire of classical model theory is an important step. For not only is this step necessary to make sure that semantics—the general theory of the relationships between languages and models—be independent of any bias concerning what kinds of model should eventually be allowed; it also increases the actual range of applicability of that general theory, making it possible for specific semantic analyses to be implemented and applied without resorting to artificial regimentations in the first place.

It will of course take us far afield to address these issues in concrete detail. The only way to satisfactorily answer the question of how a theory works is to see it at work, but this goes beyond the scope of this presentation. Here I shall content myself with highlighting the conceptual turn-about opened up by liberalizing the outlook of standard semantics in the indicated way. What makes the challenge interesting is not the possibility of providing some means of explaining away the aforementioned phenomena. It is, rather, the prospect of an open-faced attitude towards them, the possibility of taking them for what they are without the fear of logical disaster. In this sense, the general philosophical idea behind the account developed here is that ambiguity, fugacity of reference, inconsistency are all "traits of verbal form"[89] and do not necessarily extend to the ontology. In itself, the world may well

---

[89] The expression is Quine's [1960], p. 88.

be sharp, composed of exact structures. It is the connection between language and the world that may be incomplete and sometimes inconsistent, linking language not to a single structure, but to a whole class of such structures. A good “universal semantics” must do justice to this fact. And I should like to think that the material presented here lends itself naturally to this task.

Appendix

This Appendix outlines the intuitive set-theoretic framework presupposed in the foregoing. My purpose is both to fix notation and terminology and to set out the main assumptions with which I have been working. I also include here a statement of some relevant facts that play a role in my arguments, though their proof is systematically omitted.

Classes. I take a *class* to be any collection of objects, called the *members* of the class. If a class is itself a member of some class, then it is a *set*; otherwise it is a *proper class*. As a rule, in the following I shall use capital letters A, B, C, etc. (possibly with subscripts or superscripts) as variables for arbitrary classes, whereas lower case letters a, b, c, etc. (also with auxiliary subscripts or superscripts) will range over sets and other class members, i.e., *individuals*.

The membership relation is symbolized by the Greek letter $\in$, so that a formula of the form $a \in A$ says that a is *a member of* (is *contained in*, or *belongs to*) A. More generally, the notation $a_1, a_2, \ldots, a_n \in A$ signifies that the elements $a_1, a_2, \ldots, a_n$ are contained in A; and $A \subseteq B$ indicates that every member of A is contained in B, i.e., that A is *included in* (or is a *subclass of*) B. If $A \subseteq B$ but not vice versa, then I may write $A \subset B$, and say that A is *properly included in* (or is a *proper subclass of*) B; on the other hand, if A and B are included in each other, then I take A to be *identical with* B, and write $A=B$. (In the case of individuals, the identity relation symbolized by = is assumed as primitive. I also use = as a sign for definitional equivalence, but there is no danger of confusion.) As customary, the formulas $a \notin A$, $A \not\subseteq B$, $A \not\subset B$, and $A \neq B$ will then express the negations of $a \in A$, $A \subseteq B$, $A \subset B$, and $A=B$, respectively, and I use the convenient notation $\{a\colon \phi\}$ to designate the *class of all elements a which satisfy* ϕ, where ϕ is any formula in the metalanguage. In this regard, an expression of the form $\{a \in A\colon \phi\}$ is to be understood as synonymous with $\{a\text{: } a\in A \text{ and } \phi\}$, and similarly for $\{A\subseteq B\text{: } \phi\}$, $\{A\subset B\text{: } \phi\}$, etc. I may also use $\{a_1, a_2,\ldots,a_n\}$ as short for $\{a\text{: } a=a_1 \text{ or } a=a_2 \text{ or} \ldots a=a_n\}$. And if $\tau(a_1, a_2,\ldots,a_n)$ is a term in the free variables $a_1, a_2,\ldots,a_n$, I generally write $\{\tau(a_1, a_2,\ldots,a_n)\text{: } \phi\}$ instead of $\{a\text{: } a=\tau(a_1, a_2, \ldots,a_n) \text{ for some } a_1, a_2,\ldots,a_n \text{ such that } \phi\}$. Further obvious abbreviations are also occasionally used, when this does not impair (or does improve) readability.

Using this notation, the following customary definitions apply: $\varnothing=\{a\text{: } a\neq a\}$ (the *empty class*); $\mathcal{U}=\{a\text{: } a=a\}$ (the *universal class*); $\wp A=\{a\text{: } a\subseteq A\}$ (the *power class* of A), $\bigcup A=\{b\text{: } b\in a \text{ for some } a\in A\}$ (the *union* of A); $\bigcap A=\{b\text{: } b\in a \text{ for every } a\in A\}$ (the *intersection* of A); $B\cup C=\{a\text{: } a\in B \text{ or } a\in C\}$ (the *join* of B and C); $B\cap C=\{a\text{: } a\in B \text{ and } a\in C\}$ (the *meet* of B and C); and $B-C=\{a\text{: } a\in B \text{ and } a\notin C\}$ (the *difference* of B and C). Note that $\bigcap\varnothing=\mathcal{U}$. For convenience, however, I always identify $\bigcap\varnothing$ with $\varnothing$ itself, unless otherwise specified.

In addition to the principles of extensionality and comprehension implicit in the definitions above, I assume that the following holds as well: $\varnothing$ is a set; if a and b are elements, then $\{a,b\}$ is a set; and if A is a set, then so are $\wp A$, $\bigcup A$, and $\bigcap A$.

Relations. I identify *relations* with classes of ordered pairs, where an *ordered pair* $\langle a,b\rangle$ (with *first coordinate* a and *second coordinate* b) is defined as the set $\{\{a\},\{a,b\}\}$. In general, the letter R is used as a variable for relations, and I tend to follow the convention of writing aRb instead of $\langle a,b\rangle\in R$. Conversely, whenever an expression of the form $x*y$ is introduced by definition (where $*$ is a single symbol), I may use $*$ as a name for the relation $\{\langle a,b\rangle\text{: } a*b\}$, so that in case x and y are elements (and not proper classes) the expression $x*y$ becomes equivalent to $\langle x,y\rangle\in *$: this applies, in particular, to the relation symbols $\in$, $\subseteq$, $\subset$, $=$, etc. introduced above).

Given a relation R, the following definitions apply: $\breve{R}=\{\langle b,a\rangle\text{: } aRb\}$ (the *converse* of R); $\mathcal{D}R=\{a\text{: } aRb \text{ for some } b\}$ (the *domain* of R); $\mathcal{R}R=\{b\text{: } aRb \text{ for some } a\}$ (the *range* of R); $\mathcal{F}R=\mathcal{D}R\cup\mathcal{R}R$ (the *field* of R); $R[a]=\{b\text{: } aRb\}$ (the *image* of a under R); and $R\restriction A=\{\langle a,b\rangle\in R\text{: } a\in A \text{ and } b\in R[a]\}$ (the *restriction* of R to A). Moreover, the *relative product* of two relations R and R' is the relation $R\circ R'=\{\langle a,b\rangle\text{: } aRc \text{ and } cR'b \text{ for some } c\}$, while the *simple product* of two classes A, B is given by $A\times B$

$=\{\langle a,b\rangle: a\in A$ and $b\in B\}$. More generally, the notation $A_1\times\ldots\times A_n$ stands for $(\ldots(A_1\times A_2)\times\ldots)\times A_n$ (if $n>1$, otherwise $A_1\times\ldots\times A_n=A_1$), and I use the abbreviation $\langle a_1,\ldots,a_n\rangle$ to denote a typical element $\langle\langle\ldots\langle a_1,a_2\rangle,\ldots\rangle,a_n\rangle$ of such a product, to be called an *n-tuple of elements*. We may also define the *ordered couple* of two classes A and B (in this order) to be the unique relation $(A\times\{\varnothing\})\cup(B\times\{\{\varnothing\}\})$: this notion, for which I use the notation (A,B), provides a convenient analogue of the concept of an ordered pair, which could only be defined with respect to elements. Again, I then let $(A_1,\ldots,A_n)=((\ldots(A_1,A_2),\ldots),A_n)$ and speak of this as an *n-tuple of classes*.

Further terminology is given by the following standard definitions, where R is an arbitrary relation. Let A be a class and b an element: then b is *R-least* (or *$\breve{R}$-greatest*) in A iff $b\in A$ and $A\cap R[b]=A$; b is *R-minimal* (or *$\breve{R}$-maximal*) in A iff $b\in A$ and $A\cap\breve{R}[b]\subseteq\{b\}$; b is an *R-lower bound* (or an *$\breve{R}$-upper bound*) for $B\subseteq A$ iff $B\cap R[b]=B$; and b is the *R-infimum* (or the *$\breve{R}$-supremum*) for $B\subseteq A$, denoted by $\sqcap_R B$ (or $\sqcup_{\breve{R}}B$) iff b is R-greatest in the class of all R-lower bounds for B (in case B is a two-element set $\{a,c\}$, I write $a\sqcap_R c$ and $a\sqcup_R c$ instead of $\sqcap_R\{a,c\}$ and $\sqcup_R\{a,c\}$, referring to these as the *R-meet* and the *R-join* of B, respectively). Moreover, I say that R is *reflexive* in A iff $=\restriction A\subseteq(R\restriction A)$; R is *irreflexive* in A iff $(=\restriction A)\cap(R\restriction A)=\varnothing$; R is *transitive* in A iff $(R\restriction A)\circ(R\restriction A)\subseteq(R\restriction A)$; R is *symmetric* in A iff $R\restriction A=\breve{R}\restriction A$; R is *antisymmetric* in A iff $(R\restriction A)\cap(\breve{R}\restriction A)\subseteq(=\restriction A)$; R is *well founded* on A iff every $B\in\wp A-\{\varnothing\}$ contains an R-minimal element and R is irreflexive in A; R is a *pre-ordering* of A iff R is reflexive and transitive in A; R is an *equivalence* on A iff R is a symmetric pre-ordering of A; R is *a (partial) ordering* of A iff R is an antisymmetric pre-ordering of A; and R is a *well ordering* of A iff R is an ordering of A and every $B\in\wp A-\{\varnothing\}$ contains an R-least element. I may also refer to a couple (A,R) as a *(well) ordered class*, meaning that R is a (well) ordering of A. In particular, an ordered class (A,R) such that aRc or cRa for every $a,c\in A$ is also called a *chain* (under R), whereas it is called a *lattice* (under R) if $a\sqcap_R c\in A$ and $a\sqcup_R c\in A$ for every such a and c.

As postulates, I then assume that the relation $\in$ is well founded on every set (axiom of regularity) and that every set can be well ordered by some relation (axiom of choice). This latter postulate, in turn, implies that if (A,R) is an ordered class with the property that every chain (B,R)

with $B \subseteq A$ has an R-upper bound in A, then A contains an R-maximal element (Zorn's Lemma).

Functions. By a *function* I always understand a many-one relation: that is, a relation f such that $f[a]$ contains exactly one element for each argument $a \in \mathcal{D}f$. (I generally use the letters f, g, and h as special variables for functions.) As customary, the unique element of $f[a]$ is denoted by $f(a)$ (sometimes by f_a, or f^a), while $f(a/z)$ denotes that function which is obtained from f by replacing $\langle a, f(a)\rangle$ (if defined) with $\langle a,z\rangle$. Moreover, if $\tau(a)$ is any expression that denotes an element for each element a satisfying a condition ϕ, I may write $\langle \tau(a){:}\ \phi\rangle$ for $\{\langle a,\tau(a)\rangle{:}\ \phi\}$, using τ itself as a name for this function. Occasionally, when ϕ is of the particular form $a \in A$ I may also speak of $\langle \tau(a){:}\ \phi\rangle$, not as a function with domain A, but as a *system of elements indexed* by A (or an A-*termed sequence*), referring to $\tau(a)$ as the *a-th term* of the system (sequence): the difference is purely verbal, but the use of this terminology is to suggest that $\langle \tau(a){:}\ \phi\rangle$ is intuitively identified with its range, i.e., with the *indexed class* $\{\tau(a){:}\ a \in A\}$].

A function f is also called a *mapping* of A *into* B, written $f{:}\ A \to B$, iff $\mathcal{D}f = A$ and $\mathcal{R}f \subseteq B$. In particular, f is *onto* B ($f{:}\ A \twoheadrightarrow B$) iff $\mathcal{R}f = B$; f is *one-one* ($f{:}\ A \rightarrowtail B$) iff $\breve{f}$ is also a function; and f is a *correspondence* between A and B ($f{:}\ A \rightarrowtail\!\!\!\!\twoheadrightarrow B$) iff f is one-one and onto B. In case there is a one-one mapping of A into B I may also $A \lesssim B$ or $B \gtrsim A$, and I may write $A \approx B$ in case the mapping is a correspondence. Moreover, I write B^A for $\{f{:}\ f{:}\ A \to B\}$ (the A-*th Cartesian power* of B) and $\prod\langle \tau(a){:}\ \phi\rangle$ for $\{f{:}\ f{:}\ \mathcal{D}\tau \to \bigcup \mathcal{R}\tau$ so that $f(a) \in \tau(a)$ for every a such that $\phi\}$ (the *Cartesian product* of $\langle \tau(a){:}\ \phi\rangle$). Thus, $B^A = \prod\langle \tau(a){:}\ a \in A\rangle$ whenever B is a set and $\mathcal{R}\tau = \{B\}$. However, if A is a natural number $m \neq 0$ (see below), I systematically identify B^A with $B \times \ldots \times B$ (m times), unless otherwise specified. In this case, a subclass $R \subseteq B^A$ may be referred to as an *m-ary relation* on B, and as an *(m–1)-ary operation* on B if, in addition, R is a mapping of B^{m-1} into B (where $m-1 \neq 0$): the notation $R[b_1,\ldots,b_m]$ can then be used as an abbreviation for $R[\langle b_1,\ldots,b_m\rangle]$, and if f is a function, $f(b_1,\ldots,b_m)$ is the unique element of $f[b_1,\ldots,b_m]$. In this regard, I also say that a subclass $B' \subseteq B$ is *closed under* an m-ary relation R on B if $R[b_1,\ldots,b_m] \subseteq B'$ for every $b_1, \ldots, b_m \in B'$: accordingly, the *closure of* B' *under* R (denoted by $Cl_R B'$) is the $\subseteq$-least class $C \subseteq B$ including B' and closed under R, and R is *well-grounded* on B' iff $R[b_1,\ldots,b_m] \cap B' = \varnothing$ for all elements $b_1,\ldots,b_m \in Cl_R B'$. (*Par abus de langage*, the same terminology applies also when R is a mapping of $(\wp B - \{\varnothing\})^m$ into B and $b_1,\ldots,b_m$ range over non empty sets.)

As a general postulate on functions, I then assume that if f is a function and $\mathcal{D}f$ is a set, then $\mathcal{R}f$ is also a set (axiom of replacement). I also make use of the fact that if R is a relation and $\mathcal{D}R$ is a set, then R includes a function f such that $f(a) \in R[a]$ for all $a \in \mathcal{D}R$. This is just a variant of the axiom of choice, which I already mentioned above. Furthermore, I occasionally rely on the following form of the recursion theorem, which follows from the axioms already assumed: let A and B be sets, $f\colon A \to A$, and $C = Cl_g B$, where g is a one-one function and $\mathcal{R}g \cap B = \varnothing$; then any function $h_0\colon B \to A$ can be extended to a function $h\colon C \to A$ such that, for all $c \in C$, $h(g(c)) = f(h(c))$. Such an h is called a *homomorphism* from C into A and is uniquely defined.

Ordinals. Finally, let me recall that the notion of an *ordinal number* (or *ordinal* for short) may be defined in such a way that each ordinal coincides with the class of its predecessors. More precisely, an ordinal I understand to be a set α well ordered by the relation $\in \cup =$ and such that such that $\bigcup\alpha \subseteq \alpha$. (I generally use lower case Greek letters α, β, γ, etc. as special variables for ordinal numbers.) Thus, every ordinal is a set of ordinals; and if α and β are ordinals, then $\alpha < \beta$ (or $\beta > \alpha$) means the same as $\alpha \in \beta$, and $\alpha \leq \beta$ (or $\beta \geq \alpha$) the same as $\alpha \in \beta \cup \{\alpha\}$.

According to this definition, an ordinal number α is a *limit ordinal* if $\bigcup\alpha = \alpha$, and a *successor ordinal* otherwise. As a postulate, a limit ordinal different from $\varnothing$ is assumed to exist (axiom of infinity): the $\in$-least such ordinal is denoted by ω, and I identify the *natural numbers* with the members of ω. These are designated by the numerals 0, 1, 2, etc. (in the obvious order) or by the letters i, j, k, m, n, etc. (as variables). Moreover, if A is any set and α is the $\in$-least ordinal equinumerous with A (i.e., such that $\alpha \approx A$), then α is called the *cardinality* of A and may be denoted by $|A|$. An ordinal is then said to be a *cardinal number* (or simply a *cardinal*) iff it coincides with its own cardinality; and a set A is rated *finite*, *countable*, or *denumerable* dependent on whether $|A| < \omega$, $|A| \leq \omega$, or $|A| = \omega$, respectively (the existence of $|A|$ for every set A follows from the axiom of choice).

As for the arithmetic, I only need to make clear that the *sum* $\alpha+\beta$ of two ordinals α and β is defined as the unique ordinal $\gamma\geq\alpha$ for which one can fix a correspondence f: $(\gamma-\alpha)\rightarrowtail\!\!\twoheadrightarrow\beta$ such that, for all $\xi,\zeta\in\alpha$, $\xi\subseteq\zeta$ iff $f(\xi)\subseteq f(\zeta)$. Intuitively, this is the list of all ordinals in α followed by the list of all ordinals in β: if x is an α-termed sequence and y a β-termed sequence, we can therefore define the *concatenation* $x^\frown y$ of x and y to be that $(\alpha+\beta)$-termed sequence which, intuitively, is obtained by first listing the x_ξ's, $\xi<\alpha$, and then listing the y_ζ's, $\zeta<\beta$ — that is, $x^\frown y$ is defined by (i) $x^\frown y(\xi)=x(\xi)$ for $\xi<\alpha$, and (ii) $x^\frown y(\alpha+\zeta)=y(\zeta)$ for $\zeta<\beta$. Note, however, that when x (or y) is not a sequence, the same notation $x^\frown y$ may still be used on the understanding that x (or y) be identified with the 1-termed sequence $\{\langle 0,x\rangle\}$ (or $\{\langle 0,y\rangle\}$).

I conclude by noting that the following principle of *transfinite induction* is a simple consequence of the postulates which I am assuming: if ϕ is any property of ordinals, and if $\phi(\beta)$ holds whenever $\phi(\gamma)$ holds for every $\gamma<\beta$, then $\phi(\alpha)$ holds for every ordinal α. Moreover, the following fact allows us to introduce definitions by *transfinite recursion*: if α is any ordinal and f: $\bigcup\{A^\beta\colon \beta<\alpha\}\rightarrow A$ (where A^β is used here to denote the β-th Cartesian power of A), then there exists a unique function f' such that $\mathcal{D}f'=\alpha$ and $f'(\beta)=f(f'\upharpoonright\beta)$ for every $\beta<\alpha$.

References

This Bibliography lists only items referred to in the text. Unless otherwise specified, all entries are understood to be first editions and no indication of later printings is supplied [except for English translations].

Ajdukiewicz, K., 1935, 'Die syntaktische Konnexität', *Studia Philosophica* 1, 1–27. [Eng. trans. as 'Syntactic Connexion', in S. McCall (ed.), *Polish Logic 1920–1939*, Oxford: Clarendon Press, 1967, pp. 207–231.]

Anderson, A. R., Belnap, N. D. Jr., and Dunn, J. M., 1992, *Entailment. The Logic of Relevance and Necessity. Volume II*, Princeton: Princeton University Press.

Arruda, A. I., and Alves, E. H., 1979, 'A Semantical Study of Some Systems of Vagueness Logic', *Bulletin of the Section of Logic* 8, 139–144.

Asenjo, F. G., 1966, 'A Calculus of Antinomies', *Notre Dame Journal of Formal Logic* 7, 103–105.

Bach, E., 1984, 'Some Generalizations of Categorial Grammars', in F. Landman and F. Veltman (eds.), *Varieties of Formal Semantics. Proceedings of the Fourth Amsterdam Colloquium*, Dordrecht: Foris Publications, pp. 1–23.

Barba Escribà, J. L., 1987, *Modelos de Kripke para Semántica Supervaluacional*, Dissertation, Universidad Autónoma de Madrid.

— 1989, 'A Modal Version of Free Logic', *Topoi* 8, 131–135.

— 1993, 'Supervaluational Free Logic and the Logic of Information Growth', ms.

Bar-Hillel, Y., 1950, 'On Syntactical Categories', *The Journal of Symbolic Logic* 15, 1–15.

Bar-Hillel, Y., Gaifman, C., and Shamir, E., 1960, 'On Categorial and Phrase-Structure Grammars', *Bulletin of the Research Council of Israel* 3, 1–16.

Barwise, J., and Etchemendy, J., 1987, *The Liar. An Essay on Truth and Circularity*, New York and Oxford: Oxford University Press.

Barwise, J., and Perry, J., 1983, *Situations and Attitudes*, Cambridge (MA) and London: MIT Press.

Beaver, D. I., 1997, 'Presupposition', in J. van Benthem and A. Ter Meulen (eds.), *Handbook of Logic and Language,* Cambridge (MA): MIT Press; Amsterdam: Elsevier Science, pp. 939–1008.

Belnap, N. D. Jr., 1977, 'A Useful Four-Valued Logic', in J. M. Dunn and G. Epstein (eds.), *Modern Uses of Multiple-Valued Logics,* Dordrecht: Reidel, pp. 8–37.

Bencivenga, E., 1977, *Foundations of Free Logic*, Ph.D. Thesis, University of Toronto.

— 1978, 'Free Semantics for Indefinite Descriptions', *Journal of Philosophical Logic* 7, 389–405.

— 1980a, 'Truth, Correspondence, and Non-Denoting Singular Terms', *Philosophia* 9, 219–229.

— 1980b, 'Free Semantics for Definite Descriptions', *Logique et Analyse* 23, 393–405.

— 1980c, *Una logica dei termini singolari*, Torino: Boringhieri.

— 1981, 'Free Semantics', in M. L. Dalla Chiara (ed.), *Italian Studies in the Philosophy of Science*, Dordrecht: Reidel, pp. 31–48.

— 1983, 'Compactness of a Supervaluational Language', *The Journal of Symbolic Logic* 48, 384–386.

— 1984, 'Supervaluations and Theories', *Gräzer philosophische Studien* 28, 89–98.

— 1986, 'Free Logics', in D. Gabbay and F. Guenthner (eds.), *Handbook of Philosophical Logic, III: Non-Classical Logics*, Dordrecht: Reidel, pp. 373–426.

— 1990, 'Free from What?', *Erkenntnis* 33, 9–21.

— 1997, 'Supervaluations and Logical Form', ms.

Bencivenga, E., Lambert, K., and Van Fraassen, B. C., 1986, *Logic, Bivalence and Denotation*, Atascadero: Ridgeview.

Bergmann, M., 1977, 'Logic and Sortal Incorrectness', *Review of Metaphysics* 31, 61–70.

Blamey, S., 1986, 'Partial Logic', in D. Gabbay and F. Guenthner (eds.), *Handbook of Philosophical Logic, III: Non-Classical Logics*, Dordrecht: Reidel, pp. 1–70.

Blinov, A., 1994, 'Semantic Games with Chance Moves', *Synthese* 99, 311–327.

Bochvar, D. A., 1939, 'Od odnom tréhznačnom isčislénii i égo priménénii k analizu paradoksov klassičéskogo rasširénnogo funkcional'nogo isčislenija', *Matématičéskij Sbornik* 4, 287–308. [Eng. trans. as 'On a Three-Valued Logical Calculus and Its Application to the Analysis of the Classical Extended Functional Calculus', *History and Philosophy of Logic* 2, 1981, 87–112.]

Brink, C., 1987, 'Weak Entailment and Three-Valued Logic', *The Journal of Non-Classical Logic* 4, 5–22.

Burgess, J. A., 1986, 'The Truth Is Never Simple', *The Journal of Symbolic Logic* 51, 663–681.

— 1997, 'Supervaluations and the Propositional Attitude Constraint', *Journal of Philosophical Logic* 26, 103–119.

Burgess, J. A., and Humberstone, I. L., 1987, 'Natural Deduction Rules for a Logic of Vagueness', *Erkenntnis* 27, 197–229.

Burns, L. C., 1991, *Vagueness. An Investigation into Natural Languages and the Sorites Paradox*, Dordrecht, Boston, and London: Kluwer Academic Publishers.

Cantini, A., 1990, 'A Theory of Formal Truth Arithmetically Equivalent to ID_1', *The Journal of Symbolic Logic* 55, 244–259.

— 1996, *Logical Frameworks for Truth and Abstraction. An Axiomatic Study*, Amsterdam: Elsevier.

Carnap, R., 1934, *Logische Syntax der Sprache*, Vienna, Springer-Verlag (expanded Eng. trans. as *The Logical Syntax of Language*, London: Routledge and Kegan Paul, 1937).

— 1947, *Meaning and Necessity*, Chicago: University of Chicago Press.

Carpenter, B., 1996, *Type-Logical Semantics*, Cambridge (MA) and London: MIT Press.

Casadio, C., 1988, 'Semantic Categories and the Development of Categorial Grammars', in R. Oehrle, E. Bach and D. Wheeler (eds.), *Categorial Grammars and Natural Language Structures*, Dordrecht: Reidel, pp. 95–123.

Chang, C. C., and Keisler, H. J., 1973, *Model Theory*, Amsterdam: North-Holland.

Chierchia, G., and McConnell-Ginet, S., 1991, *Meaning and Grammar. An Introduction to Semantics*, Cambridge (MA) and London: MIT Press.

Church, A., 1940, 'A Formulation of the Simple Theory of Types', *The Journal of Symbolic Logic* 5, 55–68.

— 1941, *The Calculi of Lambda Conversion*, Princeton: Princeton University Press.

— 1951, 'A Formulation of the Logic of Sense and Denotation', in P. Henle, H. Kallen, and S. K. Langer (eds.), *Structure, Method and Meaning. Essays in Honour of Henry M. Sheffer*, New York: Liberal Arts.

Cleave, J. P., 1974, 'Logical Consequence in the Logic of Inexact Predicates', *Zeitschrift für mathematische Logik und Grundlagen der Mathematik* 20, 307–324.

Collins, J., and Varzi, A. C., 1997, 'Unsharpenable Vagueness', *Conference on Methods in Philosophy and the Sciences*, New York, December 6.

Cresswell, M. J., 1973, *Logics and Languages*, London: Methuen.
— 1977, 'Categorial Languages', *Studia Logica* 36, 257–269.
Curry, H. B., 1929, 'An Analysis of Logical Substitution', *American Journal of Mathematics* 51, 363–384.
— 1930, 'Grundlagen der kombinatorischen Logik', *American Journal of Mathematics* 52, 509–536, 789–834.
— 1953, 'Les Systèmes Formels et les Langues', *Colloques Internationaux du Centre National de la Recherche Scientifique* 36, 1–10.
Da Costa, N. C. A., and Alves, E. H., 1976, 'Une sémantique pour le calcul C_1', *Comptes Rendus Hebdomadaires des Séances de l'Académie des Sciences de Paris* 283A, 729–731.
Da Costa, N. C. A., and Dubikajtis, L., 1977, 'On Jaśkowski's Discussive Logic', in A. I. Arruda, N. C. A. Da Costa, and R. Chuaqui (eds.), *Non-Classical Logic, Model Theory and Computability*, Amsterdam: North Holland, pp. 37–56.
Day, T. J., 1992, 'Excluded Middle and Bivalence', *Erkenntnis* 37, 93–97.
Dienes, Z. P., 1949, 'On an Implication Function in Many-Valued Systems of Logic', *The Journal of Symbolic Logic* 14, 95–97.
Dummett, M., 1975, 'Wang's Paradox', *Synthese* 30, 301–324.
— 1991, *The Logical Basis of Metaphysics*, Cambridge (MA): Harvand University Press,
— 1995, 'Bivalence and Vagueness', *Theoria* 61, 201–216.
Dunn, J. M., 1976, 'Intuitive Semantics for First-Degree Entailments and "Coupled Trees"', *Philosophcal Studies* 29, 149–168.
— 1997, 'Partiality and Its Dual', *First World Congress on Paraconsistency*, Gent (Belgium), August 2.
Etchemendy, J., 1990, *The Concept of Logicl Consequence*, Cambridge: Harvard University Press.
Fagin, R., and Halpern, J. Y., 1987, 'Belief, Awareness, and Limited Reasoning', *Artificial Intelligence* 34, 39–76.
Fagin, R., Halpern, J. Y., Moses, Y., and Vardi, M. Y., 1995, *Reasoning About Knowledge*, Cambridge (MA) and London: MIT Press.
Feferman, S., 1984, 'Toward Useful Type-Free Theories, I', *The Journal of Symbolic Logic* 49, 75–111.
Fenstad, J. E., 1997, 'Partiality', in J. van Benthem and A. Ter Meulen (eds.), *Handbook of Logic and Language,* Cambridge (MA): MIT Press; Amsterdam: Elsevier Science, pp. 649–682.
Fine, K., 1975, 'Vagueness, Truth and Logic', *Synthese* 30, 265–300.
— 1983, 'The Permutation Principle in Quantificational Logic', *Journal of Philosophical Logic* 12, 33–37.
Fitting, M., 1986, 'Notes on the Mathematical Aspects of Kripke's Theory of Truth', *Notre Dame Journal of Formal Logic* 27, 75–88.

— 1987, 'Partial Models and Logic Programming', *Theoretical Computer Science* 48, 229–255.
— 1989, 'Bilattices and the Theory of Truth', *Journal of Philosophical Logic* 18, 225–256.
— 1992, 'Kleene's Logic, Generalized', *Journal of Logic and Computation* 1, 797–810.
Fodor, J. A., and LePore, E., 1996, 'What Cannot Be Evaluated Cannot Be Evaluated, and It Cannot Be Supervalued Either', *The Journal of Philosophy* 93, 516–535.
Frege, G., 1892, 'Über Sinn und Bedeutung', *Zeitschrift für Philosophie und philosophische Kritik (N.S.)* 100, 25–50. [Eng. trans. as 'On Sense and Reference', in *Translations from the Philosophical Writings of Gottlob Frege*, ed. by P. T. Geach and M. Black, Oxford: Backwell, 1952, pp. 56–78.]
— 1893, *Grundgesetze der Arithmetik, begriffsschriftlich abgeleitet*, Band I, Jena, Pohle (partial Eng. trans. as *The Basic Laws of Arithmetic: Exposition of the System*, Berkeley and Los Angeles: University of California Press, 1964).
— 1903, *Grundgesetze der Arithmetik, begriffsschriftlich abgeleitet*, Band II, Jena, Pohle (partial Eng. trans. in *Translations from the Philosophical Writings of Gottlob Frege*, ed. by P. T. Geach and M. Black, Oxford: Backwell, 1952, pp. 21–41).
Fuhrmann, A., 1997, *An Essay on Contraction*, Stanford (CA): CSLI Publications.
García-Carpintero, M., 1993, 'The Grounds for the Model-Theoretic Account of the Logical Properties', *Notre Dame Journal of Formal Logic* 34, 107–131.
Gärdenfors, P., 1988, *Knowledge in Flux. Modeling the Dynamics of Epistemic States*, Cambridge (MA) and London: MIT Press.
Garson, J. W., 1984, 'Quantification in Modal Logic', in D. Gabbay and F. Guenthner (eds.), *Handbook of Philosophical Logic, II: Extensions of Classical Logic*, Dordrecht: Reidel, pp. 249–307.
Geach, P. T., 1970, 'A Program for Syntax', *Synthese* 22, 3–17.
Ginsberg, M., 1988, 'Multivalued Logics: A Uniform Approach to Reasoning in Artificial Intelligence', *Computational Intelligence* 4, 265–316.
— 1990, 'Bilattices and Modal Operators', *Journal of Logic and Computation* 1, 41–69.
Grant, J., 1974, 'Incomplete Models', *Notre Dame Journal of Formal Logic* 15, 601–607.
— 1975, 'Inconsistent and Incomplete Logics', *Mathematics Magazine* 48, 154–159.
Griffin, N., 1978, 'Supervaluations and Tarski', *Notre Dame Journal of Formal Logic* 19, 297–298.

Gupta, A., and Belnap, N. D. Jr., 1993, *The Revision Theory of Truth*, Cambridge (MA) and London: MIT Press.

Hailperin, T., 1953, 'Quantification Theory and Empty Individual-Domains', *The Journal of Symbolic Logic* 18, 197–200.

Heller, M., 1990, *The Ontology of Physical Objects: Four-Dimensional Hunks of Matter*, Cambridge, Cambridge University Press.

Henkin, L., 1950, 'The Algebraic Characterization of Quantifiers', *Fundamenta Mathematicae* 37, 63–74.

Herzberger, H. G., 1975a, 'Canonical Superlanguages', *Journal of Philosophical Logic* 4, 45–65.

— 1975b, 'Supervaluations in Two Dimensions', *Proceedings of the 1975 International Symposium on Multiple-Valued Logics*, New York: IEEE Press, pp. 429–435.

— 1975c, 'Presuppositional Policies', *Philosophia* 5, 243–268.

— 1980, 'Supervaluations without Truth-Value Gaps', *Canadian Journal of Philosophy*, suppl. vol. 6, 15–27.

— 1982, 'The Algebra of Supervaluations', *Topoi* 1, 74–81.

Hintikka, K. J. J., 1959, 'Existential Presuppositions and Existential Commitments', *The Journal of Philosophy* 56, 125–137.

Horwich, P., 1990, *Truth*, Oxford: Blackwell.

Husserl, E., 1900/1, *Logische Untersuchungen*, Halle: Max Niemeyer (2nd ed. 1913/21; Eng. trans. as *Logical Investigations*, London: Routledge and Kegan Paul, 1970).

Hyde, D., 1997, 'From Heaps and Gaps to Heaps and Gluts', *Mind* 106, 641–660.

Jackson, F., and Pargetter, R., 1988, 'A Question about Rest and Motion', *Philosophical Studies* 53, 141–146.

Jacobson, P., 1996, 'The Syntax/Semantics Interface in Categorial Grammar', in S. Lappin (ed.), *The Handbook of Contemporary Semantic Theory*, London, Blackwell, pp. 69–106.

Jaśkowski, S., 1948, 'Rachunek zdańdla systemów dedukcyjnych sprzecznych', *Studia Societatis Scientiarum Torunensis, Sectio A* 1, 8, 55–77. [Eng. trans. as 'Propositional Calculus for Contradictory Deductive Systems, *Studia Logica* 24, 1969, 143–157.]

Jennings, R. E., and Schotch, P. K., 1984, 'The Preservation of Coherence', *Studia Logica* 43, 89–106.

Kamareddine, F., and Klein, E., 1993, 'Nominalization, Predication and Type Containment', *Journal of Logic, Language and Information* 2, 171–215.

Kamp, H., 1975, 'Two Theories about Adjectves', in E. L. Keenan (ed.), *Formal Semantics and Natural Language*, Cambridge: Cambridge University Press, pp. 45–89.

— 1981a, 'The Paradox of the Heap', in E. Mönnich (ed.), *Aspects of Philosophical Logic*, Dordrecht: Reidel, pp. 225–277.

— 1981b, 'A Theory of Truth and Semantic Representation', in J. Groenedijk, T. Janssen, and M. J. Stokhof (eds.), *Formal Methods in the Study of Language*, Amsterdam: Matematisch Centrum, pp. 277–322.

Keefe, R., and Smith, P., 1997, 'Introduction: Thories of Vagueness', in R. Keefe and P. Smith (eds.), *Vagueness: A Reader*, Cambridge (MA) and London: MIT Press, pp. 1–57.

Kleene, S. K., 1938, 'On a Notation for Ordinal Numbers', *The Journal of Symbolic Logic* 3, 150–155.

Klein, E., 1980, 'A Semantics for Positive and Comparative Adjectives', *Linguistics and Philosophy* 4, 1–45.

Konolige, K., 1985, 'Belief and Incompleteness', in J. R. Hobbs and R. Moore (eds.), *Formal Theories of the Commonsense World*, Norwood: Ablex, 359–404.

Kotas, J., and Da Costa, N. C. A., 1979, 'A New Formulation of Discussive Logic', *Studia Logica* 38, 429–444.

Kripke, S., 1963, 'Semantical Considerations on Modal Logic', *Acta Philosophica Fennica* 16, 83–94.

— 1975a, 'Lectures on Truth', Princeton University, ms.

— 1975b, 'Outline of a Theory of Truth', *The Journal of Philosophy* 72, 690–716.

Kyburg, H. E., Jr., 1970, 'Conjunctivitis', in M. Swain (ed.), *Induction, Acceptance, and Rational Belief*, Dordrecht: Reidel, pp. 55–82.

— 1997, 'The Rule of Adjunction and Reasonable Inference', *The Journal of Philosophy* 94, 109–125.

Lambek, J., 1958, 'The Mathematics of Sentence Structure', *American Mathematical Monthly* 65, 154–170.

— 1961, 'On the Calculus of Syntactic Types', in R. O. Jakobson (ed.), *The Structure of Language and Its Mathematical Aspects. Proceedings of the Twelfth Symposium in Applied Mathematics*, Providence: American Mathematical Society, pp. 166–178.

Lambert, K., 1963, 'Existential Import Revisited', *Notre Dame Journal of Formal Logic* 4, 288–292.

— 1969, 'Logical Truth and Microphysics', in K. Lambert (ed.), *The Logical Way of Doing Things*, New Haven: Yale University Press, pp. 93–117.

Lambert, K., and Bencivenga, E., 1986, 'A Free Logic with Simple and Complex Predicates', *Notre Dame Journal of Formal Logic* 27, 247–256.

Lambert, K., Leblanc, H., and Meyer, R. K., 1969, 'A Liberated Version of S5', *Archiv für mathematische Logik und Grundlagenforschung* 12, 151–154.

Langholm, T., 1988, *Partiality, Truth and Persistence*, Stanford (CA): CSLI Publications.

Léa Sombé (Besnard, P., *et al.*), 1990, 'Reasoning under Incomplete Information in Artificial Intelligence: A Comparison of Formalisms Using a Single Example', *International Journal of Intelligent Systems* 5, 323–472.

Leblanc, H., and Hailperin, T., 1959, 'Nondesignating Singular Terms', *Philosophical Review* 68, 39–43.

Leblanc, H., and Meyer, R. K., 1970, 'On Prefacing $(\forall X)A \supset A(Y/X)$ with $(\forall Y)$. A Free Quantification Theory without Identity', *Zeitschrift für mathematische Logik und Grundlagen der Mathematik* 16, 447–462.

Lehmann, S., 1994, 'Strict Fregean Free Logic', *Journal of Philosophical Logic* 23, 307–336.

Leonard, H. S., 1956, 'The Logic of Existence', *Philosophical Studies* 7, 49–64.

Leśniewski, S., 1929, 'Grundzüge eines neuen Systems der Grundlagen der Mathematik', *Fundamenta Mathematicae* 14, 1–81. [Eng. trans. as 'Fundamentals of a New System of the Foundations of Mathematics', in S. Leśniewski, *Collected Works*, ed. by S. J. Surma, J. T. Srzednicki, D. I. Barnett, and V. F. Rickey, Dordrecht, Boston, and London: Kluwer Academic Publishers, 1992, Vol. 2, pp. 410–605.]

Levesque, H. J., 1984, 'A Logic of Implicit and Explicit Belief', *Proceedings of AAAI-84. Fourth National Conference on Artificial Intelligence*, Austin: AAAI [William Kaufmann], pp. 198–202.

— 1990, 'All I Know: A Study in Autoepistemic Logic',*Artificial Intelligence* 42, 263–309.

Levi, I., 1980, *The Enterprise of Knowledge*, Cambridge (MA) and London: MIT Press.

— 1991, *The Fixation of Belief and Its Undoing*, Cambridge: Cambridge University Press.

Levin, H. D., 1982, *Categorial Grammar and the Logical Form of Quantification*, Napoli: Bibliopolis.

Lewis, D. K., 1970, 'General Semantics', *Synthese* 22, 18–67.

— 1978, 'Truth in Fiction',*American Philosophical Quarterly* 15, 37–46.

— 1979, 'Scorekeeping in a Language Game', *Journal of Philosophical Logic* 8, 339–359.

— 1982, 'Logic for Equivocators', *Noûs* 16, 431–441.

— 1983a, 'Postscripts to "General Semantics"', in *Philosophical Papers. Volume 1*, New York and Oxford: Oxford University Press, pp. 230–232.

— 1983b, 'Postscripts to "Truth in Fiction"', in *Philosophical Papers, Volume 1*, New York and Oxford: Oxford University Press, pp. 276–280.

Łukasiewicz, J., 1920, 'O logice trójwartościowej', *Ruch Filozoficzny* 5, 169–171. [Eng. trans. as 'On Three-Valued Logic', in S. McCall (ed.), *Polish Logic 1920–1939*, Oxford: Clarendon, 1967, pp. 16–18.]

Lyons, J., 1969, *Introduction to Theoretical Linguistics*, Cambridge: Cambridge University Press.

Machina, K., 1976, 'Truth, Belief, and Vagueness', *Journal of Philosophical Logic* 5, 47–78.

Makinson, D. C., 1964, 'The Paradox of the Preface', *Analysis* 25, 205–207.

Marconi, D., 1981, 'Types of Non-Scotian Logic', *Logique et Analyse* 95/96, 407–414.

Marsh, W., and Partee, B. H., 1987, 'How Non-Context Free is Variable Binding?', in W. J. Savitch, E. Bach, W. Marsh, and G. Safran-Naveh (eds.), *The Formal Complexity of Natural Language*, Dordrecht: Reidel, pp. 369–386.

Martin, J. N., 1984, 'Epistemic Semantics for Classical and Intuitionistic Logic', *Notre Dame Journal of Formal Logic* 25, 105–116.

Martin, R. L., 1970, 'A Category Solution to the Liar', in R. L. Martin (ed.), *The Paradox of the Liar*, Atascadero (CA): Ridgeview, pp. 91–112.

Martin, R. L., and Woodruff, P. W., 1975, 'On Representing 'True-in-*L*'in *L*', *Philosophia* 5, 213–217.

Martin-Löf, P., 1984, *Intuitionistic Type Theory*, Neaples: Bibliopolis.

McCall, S., 1966, 'Excluded Middle, Bivalence, and Fatalism', *Inquiry* 4, 384–386.

McGee, V., 1991, *Truth, Vagueness, and Paradox. An Essay in the Logic of Truth*, Indianapolis and Cambridge (MA): Hackett.

McGee, V., and McLaughlin, B., 1994, 'Distinctions Without a Difference', *The Southern Journal of Philosophy*, 33 (Suppl.), 203–251.

Mehlberg, H., 1956, *The Reach of Science*, Toronto: Toronto University Press.

Meinong, A., 1904, 'Über Gegenstandstheorie', in A. Meinong, R. Ameseder, and E. Mally (eds.), *Untersuchungen zur Gegenstandstheorie und Psychologie*, Leipzig: Barth, pp. 1–50. [Eng. trans. as 'On The Theory of Objects', in R. M. Chisholm (ed.), *Realism and the Background of Phenomenology*, New York: The Free Press, 1960, pp. 76–117]

Meyer, R. K., Bencivenga, E., and Lambert, K., 1982, 'The Ineliminability of E! in Free Quantification Theory Without Identity', *Journal of Philosophical Logic* 11, 229–231.

Meyer, R. K., and Lambert, K., 1968, 'Universally Free Logic and Standard Quantification Theory', *The Journal of Symbolic Logic* 33, 8–26.

Mikenberg, I., Da Costa, N., and Chuaqui, R., 1986, 'Pragmatic Truth and Approximation to Truth', *Journal of Symbolic Logic* 51, 201–221.

Montague, R., 1970a, 'Universal Grammar', *Theoria* 36, 373–398.

— 1970b, 'English as a Formal Language', in B. Visentini *et al.* (eds.), *Linguaggi nella società e nella tecnica*, Milano: Edizioni di Comunità, pp. 189–224.

Morgan, C. G., 1970, 'Weak Liberated Versions of T and S4', *The Journal of Symbolic Logic* 40, 25–36.

Morreau, M., 1998, 'Supervaluations Can Have Truth-Value Gaps After All', *Conference on Vagueness*, Bled (Slovenia), June 6.

Mostowski, A., 1951, 'On the Rules of Proof in the Pure Functional Calculus of the First Order', *The Journal of Symbolic Logic* 16, 107–111.

Muskens, R., 1989, 'Going Partial in Montague Semantics', in R. Bartsch, J. van Benthem, P. van Emde Boas (eds.), *Semantics and Contextual Expression*, Dordrecht: Foris Publications, pp. 175–220.

— 1995, *Meaning and Partiality*, Stanford (CA): CSLI Publications.

Neale, S., 1990, *Descriptions*, Cambridge (MA) and London: MIT Press.

Ostertag, O., 1998, 'Introduction', in O. Ostertag (ed.), *Definite Descriptions: A Reader*, Cambridge (MA) and London: MIT Press, pp. 1–34.

Parsons, T., 1980, *Nonexistent Objects*, New Haven and London: Yale University Press.

Parsons, T., and Woodruff, P. W, 1980, 'Wordly Indeterminacy of Identity', *Proceedings of the Aristotelian Society* 95, 171–191.

Partee, B. H., 1975, 'Montague Grammar and Transformational Grammar', *Linguistic Inquiry* 6, 203–300.

— 1976, 'Some Transformational Extensions of Montague Grammar', in B. H. Partee (ed.), *Montague Grammar*, New York: Academic Press, pp. 51–76.

Patel-Schneider, P. F., 1989, 'A Four-Valued Semantics for Terminological Logics', *Artificial Intelligence* 38, 319–351.

Peirce, C. S., 1880, 'A Boolean Algebra with One Constant', published posthumously in *Collected Papers of Charles Sanders Peirce*, ed. by C. Hartshorne and P. Weiss, Cambridge (MA): Belknap Press, 1960, Volume IV, pp. 12–20.

Peña, L., 1989, 'Verum et ens convertuntur', in G. Priest, R. Routley, and J. Norman (eds.), *Paraconsistent Logic. Essays on the Inconsistent*, München: Philosophia-Verlag, pp. 563–612.

Pinkal, M., 1983, 'Towards a Semantics of Precization', in T. Ballmer and M. Pinkal (eds.), *Approaching Vagueness*, Amsterdam: North-Holland, pp. 13–57.

— 1984, 'Consistency and Context Change', in F. Landman and F. Veltman (eds.), *Varieties of Formal Semantics. Proceedings of the Fourth Amsterdam Colloquium*, Dordrecht: Foris Publications, pp. 325–342.

Priest, G., 1979, 'The Logic of Paradox', *Journal of Philosophical Logic* 8, 219–241.
— 1987, *In Contradiction. A Study of the Transconsistent*, Boston and Dordrecht: Nijhoff.
— 1989, 'Reasoning about Truth', *Artificial Intelligence*, 39, 231–244.
Priest, G., and Routley, R., 1989a, 'First Historical Introduction. A Preliminary History of Paraconsistent and Dialectic Approaches', in G. Priest, R. Routley, and J. Norman (eds.), *Paraconsistent Logic. Essays on the Inconsistent*, München: Philosophia-Verlag, pp. 3–75.
— 1989b, 'Systems of Paraconsistent Logic', in G. Priest, R. Routley, and J. Norman (eds.), *Paraconsistent Logic. Essays on the Inconsistent*, München: Philosophia-Verlag, pp. 151–186.
Przełęcki, M., 1964, 'Z semantyki pojec otwartych', *Studia Logica* 15, 189–220. [Eng. trans. as 'The Semantics of Open Concepts', in J. Pelc (ed.), *Semiotics in Poland, 1896–1969*, Dordrecht: Reidel, 1979, pp. 284–317.]
— 1969, *The Logic of Empirical Theories*, London: Routledge & Kegan Paul.
— 1976, 'Fuzziness as Multiplicity', *Erkenntnis* 10, 317–380.
— 1980, 'Set Theory as an Ontology for Semantics', in R. Haller and W. Grassl (eds.), *Language, Logic, and Philosophy. Proceedings of the Fourth International Wittgenstein Symposium*, Vienna: Hölder-Pichler-Tempsky, pp. 115–123.
— 1982, 'The Law of Excluded Middle and the Problem of Idealism', *Grazer Philosophische Studien* 18, 1–16.
Quine, W. V. O., 1939, 'Designation and Existence', *Journal of Philosophy* 36, 701–709.
— 1954, 'Quantification and the Empty Domain', *The Journal of Symbolic Logic* 19, 177–179.
— 1960, *Word and Object*, Cambridge (MA): MIT Press.
— 1970, *Philosophy of Logic*, Englewood Cliffs (NJ): Prentice-Hall.
— 1985, 'Events and Reification', in E. LePore and B. P. McLaughlin (eds.), *Actions and Events. Perspectives in the Philosophy of Donald Davidson*, Oxford: Blackwell, pp. 162–171.
Ranta, A., 1994, *Type-Theoretical Grammar*, Oxford: Clarendon Press.
Rasmussen, S. A., 1990, 'Supervaluational Anti-Realism and Logic', *Synthese* 84, 97–138.
Rescher, N., 1969, *Many-Valued Logic*, New York, McGraw-Hill.
— 1979, 'Mondi possibili non-standard', in D. Marconi (ed.), *La formalizzazione della dialettica*, Torino: Rosenberg & Sellier, pp. 354–417.
Rescher, N., and Brandom, R., 1980, *The Logic of Inconsistency. A Study in Non-Standard Possible-World Semantics and Ontology*, Oxford: Basil Blackwell.

Restall, G., 1995, 'Lukasiewicz, Supervaluations, and the Future', Australian National University, Technical Report TR-ARP-21-95.

Rolf, B., 1984, 'Sorites', *Synthese* 58, 219–250.

Roos, N., 1992, 'A Logic for Reasoning with Inconsistent Knowledge', *Artificial Intelligence* 57, 69–103.

Routley, R., 1980, *Exploring Meinong's Jungle*, Monograph. N. 3, Philosophy Department, Australian National University.

Routley, R., and Routley, V., 1972, 'Semantics of First Degree Entailment', *Noûs* 6, 335–359.

Rozeboom, W. W., 1962, 'The Factual Content of Theoretical Concepts', in H. Feigl and G. Maxwell (eds.), *Minnesota Studies in the Philosophy of Science, Volume III: Scientific Explanation, Space, and Time*, Minneapolis: University of Minnesota Press, pp. 273–357.

Russell, B. A. W., 1905, 'On Denoting', *Mind* 14, 479–493.

— 1908, 'Mathematical Logic as Based on the Theory of Types', *American Journal of Mathematics* 30, 222–262.

Sanchis, L. E., 1964, 'Types in Combinatory Logic', *Notre Dame Journal of Formal Logic* 5, 161–180.

Sanford, D. H., 1976, 'Competing Semantics of Vagueness: Many Values Versus Super-Truth', *Synthese* 33, 195–210.

Sayward, C., 1989, 'Does the Law of Excluded Middle Require Bivalence?', *Erkenntnis* 31, 129–137.

Schock, R., 1964, 'Contributions to Syntax, Semantics and the Philosophy of Science', *Notre Dame Journal of Formal Logic* 5, 241–289.

Schönfinkel, M., 1924, 'Über die Bausteine der mathematischen Logik', *Mathematische Annalen* 92, 305–316. [Eng. trans. as 'On the Building Blocks of Mathematical Logic', in J. van Heijenoort (ed.), *From Frege to Gödel: A Sourcebook in Mathematical Logic, 1879–1931*, Cambridge (MA): Harvard University Press, 1967, pp. 355–366.]

Schotch, P. K., and Jennings, R. E., 1989, 'On Detonating', in G. Priest, R. Routley, and J. Norman (eds.), *Paraconsistent Logic. Essays on the Inconsistent*, München: Philosophia-Verlag, pp. 306–327.

Scott, D., 1963, 'Existence and Descriptions in Formal Logic', in R. Schoenman (ed.), *Bertrand Russell, Philosopher of the Century*, London: Allen & Unwin, pp. 181–200.

— 1972, 'Continuous Lattices', in F. W. Lawvere (ed.), *Toposes, Algebraic Geometry and Logic. Partial Report on a Conference on Connections Between Category Theory and Algebraic Geometry and Intuitionistic Logic*, Berlin and Heidelberg: Springer-Verlag, pp. 97–136.

Sheffer, H. M., 1913, 'A Set of Five Independent Postulates for Boolean Algebras with Application to Logical Constants', *Transactions of the American Mathematical Society* 14, 481–488.

Sher, G., 1991, *The Bounds of Logic. A Generalized Viewpoint*, Cambridge (MA) and London: MIT Press.

— 1999, 'Is Logic a Theory of The Obvious?', in A. C. Varzi (ed.), *The Nature of Logic*, Stanford (CA): CSLI Publications, in press.

Silvestrini, D., 1981, 'Alcune considerazioni sul metodo delle supervalutazioni e una semantica a supervalutazioni per il calcolo predicativo classico', in S. Bernini (ed.), *Atti del Congresso Nazionale di Logica*, Napoli: Bibliopolis, pp. 335–350.

Simons, P., 1992, 'Vagueness and Ignorance', *Proceedings of the Aristotelian Society*, Suppl. 66, 163–177.

Skyrms, B., 1968, 'Supervaluations: Identity, Existence, and Individual Concepts', *The Journal of Philosophy* 69, 477–482.

Smiley, T., 1960, 'Sense Without Denotation', *Analysis* 20, 124–135.

Soames, S., 1989, 'Presupposition', in D. Gabbay and F. Guenthner (eds.), *Handbook of Philosophical Logic, IV: Topics in the Philosophy of Language*, Dordrecht: Reidel, pp. 553–616.

Sorensen, R., 1988, 'Precisification by Means of Vague Predicates', *Notre Dame Journal of Formal Logic* 29, 267–274.

Steedman, M., 1988, 'Combinators and Grammars', in R. Oehrle, E. Bach, and D. Wheeler (eds.), *Categorial Grammars and Natural Language Structures*, Dordrecht: Reidel, pp. 417–442.

— 1993, 'Categorial Grammar. Tutorial Overview', *Lingua* 90, 221–258.

Sternefeld, W., 1979, 'Remarks on Presuppositions, Truth, and Modality in Supervaluational Logic', *Logique et Analyse* 22, 31–47.

Stiver, J. L., 1975, 'Presupposition, Implication, and Necessitation', *Southern Journal of Philosophy* 13, 99–108.

Strawson, P. F., 1950, 'On Referring', *Mind* 69, 320–344.

— 1952, *Introduction to Logical Theory*, London: Methuen.

Tappenden, J., 1993, 'The Liar and Sorites Paradoxes: Toward a Unified Treatment', *The Journal of Philosophy* 90, 551–577.

Tarski, A., 1933, 'Pojęcie prawdy w językach nauk dedukcyjnych', *Travaux de la Société des Sciences et des Lettres de Varsovie* 4, V–116. [Eng. trans. as 'The Concept of Truth in Formalized Languages', in A. Tarski, *Logics, Semantics, Metamathematics, Papers from 1923 to 1938*, Oxford: Clarendon Press, 1956, pp. 152–278.]

— 1952, 'Some Notions and Methods on the Borderline of Algebra and Mathematics', *Proceedings of the International Congress of Mathematicians*, Providence: American Mathematical Society, pp. 705–720.

Thijsse, E. G. C., 1990, 'Partial Propositional and Modal Logic: The Overall Theory', in M. Stokhof and L. Torenvliet (eds.), *Proceedings of the Seventh Amsterdam Colloquium*, Amsterdam: ITLI, pp. 555–579.

Thomason, R. H., 1970, 'Indeterministic Time and Truth-Value Gaps', *Theoria* 36, 264–281.

— 1972, 'A Semantic Theory of Sortal Incorrectness', *Journal of Philosophical Logic* 1, 209–258.
— 1973, 'Supervaluations, the Bald Man and the Lottery', ms.
Tichy, P., 1982, 'Foundations of Partial Type Theory', *Reports on Mathematical Logic* 14, 59–72.
Turner, R., 1997, 'Types', in J. van Benthem and A. Ter Meulen (eds.), *Handbook of Logic and Language,* Cambridge (MA): MIT Press; Amsterdam: Elsevier Science, pp. 535–586.
Tye, M., 1989, 'Supervaluationism and the Law of Excluded Middle', *Analysis* 49, 141–143.
— 1990, 'Vague Objects', *Mind* 99, 535–557.
Urchs, M., 1994, 'On the Law of Excluded Contradiction', in P. Kolář and V. Svoboda (eds.), *Logica '93. Proceedings of the 7th International Symposium*, Prague, Filosofia, pp. 202–209.
Van Bendegem, J. P., 1993, 'The Logical Analysis of Time and the Problem of Indeterminism', *Communication and Cognition* 26, 209–230.
Van Benthem, J., 1990, 'Categorial Grammar and Type Theory', *Journal of Philosophical Logic* 19, 115–168.
Van Fraassen, B. C., 1966a, 'Singular Terms, Truth-Value Gaps, and Free Logic', *The Journal of Philosophy* 63, 481–495.
— 1966b, 'The Completeness of Free Logic', *Zeitschrift für mathematische Logik und Grundlagen der Mathematik* 12, 219–234.
— 1968, 'Presupposition, Implication and Self-Reference', *The Journal of Philosophy* 69, 136–152.
— 1969, 'Presuppositions, Supervaluations and Free Logic', in K. Lambert (ed.), *The Logical Way of Doing Things*, New Haven: Yale University Press, pp. 67–91.
— 1970a, 'Truth and Paradoxical Consequence', in R. L. Martin (ed.), *The Paradox of the Liar*, Atascadero (CA): Ridgeview, pp. 13–23.
— 1970b, *Formal Semantics and Logic*, New York: Macmillan.
Van Fraassen, B. C., and Lambert, K., 1967, 'On Free Description Theory', *Zeitschrift für mathematische Logik und Grundlagen der Mathematik* 13, 452–472.
Van Inwagen, P., 1988, 'How to Reason about Vague Objects', *Philosophical Topics* 16:1, 255–284.
Vardi, M. Y., 1986, 'On Epistemic Logic and Logical Omniscience', in J. Y. Halpern (ed.), *Theoretical Aspects of Reasoning about Knowledge. Proceedings of the 1986 Conference*, Los Altos (CA): Morgan Kaufmann, pp. 293–305.
Varzi, A. C., 1991, 'Truth, Falsehood and Beyond', in L. Albertazzi and R. Poli (eds.), *Topics in Philosophy and Artificial Intelligence*, Bolzano: Mitteleuropäisches Kulturinstitut, pp. 39–50.

— 1993, 'Do We Need Functional Abstraction?', in J. Czermak (ed.), *Philosophy of Mathematics. Proceedings of the 15th International Wittgenstein Symposium*, Part 1, Vienna: Hölder-Pichler-Tempsky, pp. 407–415.

— 1994a, 'Variable-Binders as Functors', in J. Wolenski and V. F. Sinisi (eds.), *The Heritage of K. Ajdukiewicz. Logic, Semantics, Epistemology*, Amsterdam and Atlanta (GA): Rodopi, pp. 329–344.

— 1994b, 'Model-Theoretic Conventionalism', in J. Hill and P. Koťátko (eds.), *Karlovy Vary Studies in Reference and Meaning*, Prague: Filosofia, pp. 406–430.

— 1994c, '*Super-Duper* Supervaluations', in T. Childers and O. Majer (eds.), *Logica '94. Proceedings of the 8th International Symposium*, Prague: Filosofia, pp. 17–40.

— 1997, 'Inconsistency Without Contradiction', *Notre Dame Journal of Formal Logic* 38, 621–639.

Vergauwen, R., 1984, 'Propositional Attitudes in Model-Theoretic Semantics', *Logique et Analyse* 27, 39–62.

Visser, A., 1984, 'Four-Valued Semantics and the Liar', *Journal of Philosophical Logic* 12, 181–212.

Von Kutschera, F., 1975, 'Partial Interpretations', in E. L. Keenan (ed.), *Formal Semantics and Natural Language*, Cambridge: Cambridge University Press, pp. 156–174.

Whitehead, A. N., and Russell, B. A. W., 1910, *Principia Mathematica*, Volume I, London: Cambridge University Press.

Williamson, T., 1992, 'Vagueness and Ignorance', *Proceedings of the Aristitelian Society*, Suppl. 66, 145–162.

— 1994, *Vagueness*, London and New York: Routledge.

Wittgenstein, L. J., 1913, Letter to Russell, in *Notebooks 1914–1916*, 2nd edition, ed. by G. H. von Wright and G. E. M. Anscombe, Oxford: Basil Blackwell, 1979, 127–129,

— 1956, *Remarks on the Foundations of Mathematics / Bemerkungen über die Grundlagen der Mathematik*, ed. by G. H. von Wright, R. Rhees, and G. E. M. Anscombe, Oxford: Basil Blackwell.

Wood, M. M., 1994, *Categorial Grammars*, London and New York: Routledge.

Woodruff, P. W., 1970, 'Logic and Truth-Value Gaps', in K. Lambert (ed.), *Philosophical Problems in Logic*, Dordrecht: Reidel, pp. 121–142.

— 1984a, 'On Supervaluations in Free Logic', *The Journal of Symbolic Logic* 49, 943–950.

— 1984b, 'Paradox, Truth and Logic, Part I: Paradox and Truth', *Journal of Philosophical Logic* 12, 213–232.

— 1991, 'Actualism, Free Logic, and First-Order Supervaluations', in W.

Spohn, B. van Fraassen, and B. Skyrms (eds.), *Existence and Explanation. Essays Presented in Honor of Karel Lambert*, Dordrecht, Boston, and London: Kluwer Academic Publishers, pp. 219–231.

Wright, C., 1995, 'The Epistemic Conception of Vagueness', *The Southern Journal of Philosophy*, Suppl. 33, 133–159.

Zemach, E. M., 1991, 'Vague Objects', *Noûs* 25, 323–340.

Index

This index lists all technical terms introduced in the text, notes and Appendix included. As a rule, entries and page numbers only refer to occurrences where the terms are first introduced or defined, though in some cases additional useful reference is also provided. The second and third parts of the index comprise a list of all authors cited in the footnotes and a list of all symbolic notation.

Subjects

Authors

Symbols